THE BEGINNING

The Beginning

A STUDY IN THE GREEK PHILOSOPHICAL APPROACH TO THE CONCEPT OF CREATION FROM ANAXIMANDER TO ST JOHN

by
ARNOLD EHRHARDT
Late Bishop Fraser Senior Lecturer in Ecclesiastical History in the University of Manchester

WITH A MEMOIR
by J. HEYWOOD THOMAS

MANCHESTER UNIVERSITY PRESS

Published by the University of Manchester at
The University Press
316–324 Oxford Road, Manchester 13

G.B. SBN 7190 0302 4

Printed in Great Britain by Butler & Tanner Ltd, Frome and London

Contents

PUBLISHER'S NOTE

Before his death, Arnold Ehrhardt completed a MS which his widow then entrusted to Dr Heywood Thomas, Reader in Theology in the University of Durham. This is here published under the editorship of Dr Heywood Thomas with the assistance of Professor F. F. Bruce, Rylands Professor in Biblical Criticism and Exegesis in the University of Manchester. Dr Heywood Thomas introduces the volume with a memoir of the author.

Arnold Ehrhardt

A MEMOIR

I remember attending a theological conference in 1954 and being very impressed by the remarkable erudition of one person in particular. He was a tall man with greying hair who spoke rather ponderously and with a slight German accent. The discussion had been on some recondite theme of early church history and—as far as I could see—was in danger of becoming lost in uncertain assertion and counter-assertion. It was then that this figure, as it were, descended on the discussion. The breadth of his scholarship was truly Erasmian and his mode of argument was most impressive. If I may use an analogy that would gladden his heart, he was like some strong and skilful boxer who throws punch after punch at his opponent and by this strong and constant attack simply demolishes the opposition. When he had finished speaking there was nothing more to say. This was Arnold Ehrhardt. His name was known to me because everyone who had any interest in theology had heard of the author of *The Apostolic Succession*, one of the most important books published in 1953. What I did not know then was that in a few years' time I would get to know this man well as a colleague and a close friend.

Arnold A. T. Ehrhardt was born in Königsberg on 14 May 1903, and he always insisted that he was a Prussian. His father was a surgeon and taught at the famous University of Königsberg. His mother was a teacher, the Christian daughter of Jewish parents—her maiden name was Rosenheim; and this Jewish ancestry was another thing to which Arnold used to refer with pride. The young lad attended school at Königsberg—the Königliches Friederiches Kollegium. This was a Pietist foundation at which the great Königsberg philosopher, Immanuel Kant, had been a pupil and Herder, the father of the German Romantic movement, a teacher. Later he studied law at the Universities of Erlangen, Bonn, Berlin and Königsberg. The twenties in Germany were far from gay. They were the lean years in which Germany was caught in the grip of inflation. Shops opened in the morning until noon and then again at three in the afternoon with new prices. The prices of commodities would be as much as a billion times their real value. Consequently Arnold's money as a young student was never enough to cover the cost of food, books and his maintenance. After paying for his lodgings he spent what little he had left on books unless there was some really pressing need such as a new pair of soles to his shoes; and in any event he was forced to find some means of obtaining food. At Bonn he earned his meals by driving the horse and cart that brought the supplies to the students' kitchen, but at Berlin he just had to do without his midday meal, successfully taking his mind off his discomfort by attending the only lecture available during the lunch hour—a lecture on Swahili! He worked as a fireman and a navvy and he economized by walking incredible distances. In 1924 he began his practical legal training which he completed in the session 1927–8 whilst also serving as Assistant Lecturer at the University of Göttingen. By this time he had gained the degree of Doctor of Law from the University of Königsberg, and he seemed to be at the beginning of a long career as an academic lawyer. He was appointed Lecturer in Roman Law in Freiburg University in 1929, and in 1932 he became the locum tenens of a Chair of Roman Law at the University of Frankfurt. Perhaps it is significant that the two Chairs of Law at the University of Frankfurt were both left unfilled for several years during this period. In the winter of 1934–5 the University of Lausanne also called upon him to act as Professor of Law there. Then in the summer of 1935 the blow fell—Arnold was forbidden to teach. The reasons for this prohibition were purely political and not, as one might expect, racist reasons which would clearly have been easy enough to provide. It was alleged that he was an undesirable influence on students. From the Government's point of view this was doubtless very true. Arnold was then, as always, popular with young people, and he was in fact teaching seditious doctrine; for he apparently maintained that judges should dispense justice and not simply apply the codified law.

Perhaps the most important event of Arnold's life altogether, and surely of this period, was his marriage in October 1930 to Edit Hahn, the sister of one of Arnold's school friends. He had met her frequently when she accompanied her brother on expeditions. They had gone skating in winter and he and she had often kicked each other more or less deliberately and so cemented a firm friendship. Later they had met at dances; and though the young lady preferred other partners for other dances, she invariably danced the waltz with Arnold. Both were interested in history and they were constantly meeting at lectures or outings. They were firm friends but nobody expected them to get married.

Ejected from Frankfurt University, Arnold and his family went first to East Prussia and then to Lörrach, German territory just outside Basle. The move gained them some peace, as it took some time for the secret police files to catch up with them. It was at this time that Arnold began to act as courier of the news from the Evangelical Church to Switzerland, taking the place of someone who had become incapable of carrying on with the work. The news came through the original courier whose little boy brought it to the Ehrhardt household when he came to play with the Ehrhardt children. Arnold's intention now was to study theology. He had long been concerned to find a clear answer to his questions about justice and he had found it increasingly difficult to give any clear guidance to those friends who, in their despair at the state of German law, sought his advice. What exactly led him to seek the answer to his difficulties in theology is hard to say. It may well have been the memory of his mother's firm faith and of her suggestion to him, when he went to University, that he become a clergyman, to which he characteristically replied that he would only preach on Matthew 23:3, which would get rather monotonous. Or it may have been Mrs Ehrhardt who suggested to him that he was, in fact, moving from the field of law to that of theology of law. Certainly, I recall him telling me once that she had thus inspired his venture into the realms of theology. But whatever the explanation his decision was to become a student of theology in Basle. So it was that Karl Barth gained a pupil and a friend. So too there was found a new courier for the underground news service—'the mad Professor on the bike' as he was known to the frontier guards after they had searched his briefcase only to find a Greek New Testament, a Syriac Grammar, a Latin text of St Augustine and an English detective novel! People as well as news got smuggled out of Germany in this way, and Arnold helped several people to escape—which he did with a readiness and an ingenuity that were typical of him. Once he took someone to a customs official's house—two gentlemen visitors for tea. They arrived by the front door but they left the house by the back door as they were able, quite literally, to step into Switzerland. Another occasion he took a roundabout route to Basle, crossing the frontier at a point where the guards had not been warned to keep an eye on his movements.

All this time the secret police were keeping a close watch on the Ehrhardts. Disguised as 'men from the Town Hall', 'currency officers' or 'rates inspectors', they paid regular visits to the Ehrhardt household, but they discovered nothing. They summoned Arnold to their office (again disguised as a Rates Collector's office), but they had no real evidence against him. Finally, they began to question a neighbour about Arnold's visits to Basle. Fortunately this neighbour warned the Ehrhardts and their legal friends advised his instant flight. So by crossing the frontier at an unlikely spot Arnold escaped to Switzerland and the family followed some six weeks later. Throughout those six weeks Mrs Ehrhardt held the fort till she and the family were also able to escape. The future was relatively secure because, quite unexpectedly, Arnold and his family had been invited by Dr G. K. Bell, Bishop of Chichester, to come to England. Almost, as it were, incidentally, he took his final examinations in Basle before leaving in the summer of 1939.

On arrival in Britain the Ehrhardts went to live in Sussex. The people of Wadhurst had rented and furnished a large house, and they invited two families on Bishop Bell's list to live there, providing for all their needs. This most agreeable arrangement lasted until May 1940 when all male enemy aliens were arrested. The following month all aliens were moved out of the south of England. Arnold was interned at Huyton, Liverpool, while his family went to live in a hostel in London. Shortly afterwards they were re-united in Cambridge, where Arnold had gained a scholarship to do research in Theology. For this he was awarded the degree of Ph.D. in 1944. At Cambridge he was received into the Church of England, and at the end of his stay there he was ordained into the Anglican ministry. He served his curacy in a Manchester parish. St James' Church, Birch-in-Rusholme, Manchester, was once the parish church of the distinguished Platt family of Platt Hall, and it was here that Arnold became curate. He quickly became very fond of Manchester and he spent many years in this part of the city. For three years he served as an Assistant in the German Department of the University while still acting as curate of St James'. Moving due north of Rusholme one comes to the parish of St Clement's Longsight, an area which is one of the less salubrious quarters of Manchester. It says much for both his charm and his toughness of character that Arnold soon won the hearts of his parishioners and his ministry was most successful. It was here that he wrote his best-known work, the book on Apostolic Succession. He had, of course, written several legal studies since 1930 (his books *Justa Causa Traditiones* (1930), *Litis Aestimatio* (1934) and *Romanistiche Studien* (1934) are deemed by Roman lawyers to be of fundamental importance even today); but his first English publication was a brief biography of Luther which he wrote in Rusholme. *The Apostolic Succession* put him amongst the most notable Anglican theologians of our day, and it was only fitting that he should have received the degree of B.D. from the University of Cambridge. Much as he loved his Longsight parish he was forced to move because his health had been threatened by a severe attack of bronchitis. So he moved out to the Lancashire countryside to the parish of St Mary's, Birch-in-Hopwood, which he served with his usual devotion for the five years from 1953 to 1958. Ever the scholar-parson, he was constantly writing articles and reviews, reading papers to conferences and attending seminars. From 1953 onwards he also served as Examining Chaplain to the Bishop of Manchester.

In 1958 Arnold's third career started, if indeed it was a third career; for, though now Bishop Fraser Lecturer in Ecclesiastical History, he remained the priest he had become. The cure of souls he had previously exercised in parishes he now continued among students in particular. He had a great affection for his pupils whom he used to call his 'babies', and I know that they all adored him. His tremendous scholarship was a byword among both his colleagues and his students. One staff seminar in particular I remember. Arnold had taken part in the discussion and was arguing about the interpretation of 'render unto Caesar what is Caesar's', maintaining that what lay behind this was the fact that the coin bore Caesar's head and this would offend the Jews with their hatred of idolatry. On leaving the meeting I was walking with another colleague, himself a very distinguished scholar, and he remarked to me, 'Really, I don't think there is anything that Ehrhardt doesn't know!' Not surprisingly the University promoted him to the grade of Senior Lecturer after he had served only three years as Lecturer. This pleased him very much, because it was evident to him as to everyone else that men who had achieved far less in the field of theological scholarship were being given positions of distinction. He was devoted to Manchester University and to the head of his Department, Professor E. Gordon Rupp. His friends speculated about his future in Manchester, and some expected that the University might one day create a chair for him. For some time, however, his health had not been good. Almost certainly as a result of the hardship he had endured in the twenties and later during his escape from Germany he was constantly suffering from a stomach ailment. Characteristically he would joke about this and he used to say when he was preparing the Gunning Lectures on the notion of the Beginning that it was reading Aristotle that really gave him stomach-ache! The first indication of the final illness came in 1962 when he collapsed in Manchester and he was admitted to the Manchester Royal Infirmary. His condition was eventually diagnosed and he was ordered to rest completely. This was for him the most unpleasant and difficult treatment, for, as he used to say, his work was his hobby. In January 1965 I returned to Manchester after six months' absence abroad, and I was amazed at the deterioration in his condition. He had aged suddenly and critically. Yet he lectured regularly during that final term until at the end of term he was taken ill; and, though he rallied, he did not recover but succumbed to a final attack on 18 March 1965. In his death theological scholarship lost a giant, and many to whom he had been a guide, guardian and friend mourned a hero.

J. Heywood Thomas

March 1967

Author's preface

The author of this book is very conscious of his inadequacy for the task which he has set himself. His sole apology for offering it to the public lies in the fact that the philosophical question of the 'beginning' has been widely neglected for centuries, especially by Christian theologians. Whilst the 'last things', the eschatological hopes and fears of humanity, and in particular of Christianity, will attract a ready audience whenever and in whatever form they may be discussed, the 'first things'—if they may be so called—seem to have caused little stirring since the time when, in 1215, the Fourth Lateran council decided that *creatio ex nihilo*, the creation of the world from nothing, was part of the Catholic faith. This is, perhaps, not surprising. For how many are there who would voluntarily take upon themselves the task of writing about 'nothing', and find a public—let alone a publisher—attracted by such a subject? The times and conditions have changed from long ago when Parmenides and the Eleatic school of Greek philosophy found it necessary to warn thinkers not to concern themselves with the *μὴ ὄν*—if it may be assumed that he meant by this expression something at least remotely related to the *nihil* of Pope Innocent III, or the 'nothing' of our daily speech. On an earlier occasion the present writer did dare to try and find similarities between Greek and Jewish thought on creation from nothing, but he is not aware of having evoked an echo in the philosophical and theological discussion of today. The philosophical question of 'beginning', but only in the last and shortest chapter also of 'the beginning', has been put in 1961 by the French philosopher Paul Levert; and after the completion of this book the author's attention was drawn to the existence of an article by K. v. Fritz about *ἀρχαί* in *Archiv für Begriffsgeschichte* IV, 1959, which however he has not seen so far.

It was a chance observation, described in the 'prelude' to this book, which brought its subject back to the author's mind, and made necessary a new approach to the problem of how to understand creation. The public might never have been made aware of the results of this enquiry, had not the University of Edinburgh invited the author to give the 'Gunning Lectures' for the year 1962, and accepted the title of this book as a suitable subject. If eventually the audience should have felt that it was lectured to on 'nothing', the author may still have achieved a modicum of success. In any case he wishes to express publicly his sincere gratitude for the opportunity of presenting his views, offered so unexpectedly and so generously to him by his Edinburgh colleagues. Thanks are also due to various friends—and more than friends—for their continued interest in the subject and its treatment by the author. At least one of them cannot escape being mentioned by name, J. Heywood Thomas, whose encouragement and advice have been of the greatest value to the author and his book.

A. E.

1 Prelude

DR FAUSTUS, ST JOHN AND PALAIPHATUS

In Goethe's *Faust*, part one, the magician, having returned from his Easter walk, sits down and begins to translate the prologue of the Fourth Gospel from 'the sacred original into his beloved German'. He begins, 'it is written: "In the beginning was the Word"; and here already I halt, who will help me on? I cannot estimate the word so highly; I need a different version'. He accordingly tries to translate the Logos first as 'reason', and then as 'power', but rejects both attempts. Finally he decides for what may well appear as a *tour de force*: 'The spirit aids me, now suddenly I see my way, and write assuredly, "in the beginning was the deed".'

When Goethe treated the 'sacred original' with such a deliberate violence he was no tiro. Long past was the Werther-period in Wetzlar; and the poet was an experienced, middle-aged civil servant and courtier in Saxe-Weimar.[1] He also possessed a far from negligible knowledge of Greek language, literature, and mysticism.[2] His version of John 1:1, therefore, should merit greater attention by theologians than it receives, not only as coming from a great genius, and leader of thought in Europe's greatest epoch, but even more so because he understood the Greek genius more deeply than many classical scholars before and after him, and yet professed frankly his bewilderment at one of the basic Christian texts.

Are we then to assume that Goethe's reaction indicates that John 1:1 is alien to Greek thought, that even the Logos is essentially non-Greek, and that the words 'in the beginning' signify no more than a mechanical

[1] According to Kuno Fischer, *Goethe-Schriften*, 7th ed., 1913, VII, 75, these verses were written in the spring of 1800.

[2] Humphrey Trevelyan, *Goethe and the Greeks*, 1941, 63 f., 113 f., has shown how deeply the Orphic hymns had influenced Goethe's thought not only during his student days in Strasbourg, but particularly in Weimar before his journey to Italy. It is true that Gottfried Herrmann's *Orphica* were only published in 1805, simultaneously with *Faust*, part 1, but Goethe had known, by Herder's advice, J. M. Gesner's *Orphica*, published 1764, and used it for his 'Urworte' and for his 'Natur-Aufsatz'. All the fragments subsequently referred to can be found there, particularly the 'Aristobulus fragment' from Eusebius *Praep. Evang.* XIII, 12 (O. Kern, *Orph. frg.*, 1922, no. 247).

copy from Genesis 1:1? It is very noticeable that the two great German commentaries by W. Bauer[1] and R. Bultmann,[2] and also the various English ones, treat the 'beginning' with a certain non-committal haste, although the repetition in John 1:2, 'the same was in the beginning with God', seems to cry out for attention; and it is no less true that both Bauer and Bultmann care only little for Greek roots of the Logos, but rush full steam into the wilderness of oriental and gnostic parallels. C. H. Dodd[3] too has discussed the Jewish origins of the Johannine prologue, i.e. its non-Hellenic background. It may be held that this is less than doing full justice to the text. For the Fourth Gospel originated, as I believe, from Asia Minor, the most Hellenized part of the Roman East, where even the Jewish Rabbis, as we learn from Justin Martyr's *Dialogue*, discoursed in their spare time about Platonic philosophy; and it found its readers at an unexpectedly early date in Egypt, at a time when even the Rabbis commissioned Aquila to produce a Greek version of the Old Testament. It was therefore written for Greeks; and since, from the days of the Renaissance, Europe may not have seen any man's mind more akin to the Greek genius than Goethe's, it may not be wise to neglect his reaction to the verse, 'in the beginning was the word'.

I believe that it is true to say that all the three alternative translations of the Logos offered by Faustus, reason, power, and action, are indeed nearer to the Greek mind than the 'word'. It is easy to illustrate from the political field the close connection in meaning between *ἀρχή* and *δύναμις*;[4] neither does it take much effort to find support from Greek philosophical sources for the alternative translation, 'in the beginning was reason' (*der Sinn* = *νοῦς*). It is perhaps improbable that Goethe was aware of the noble ancestry of this thesis, Anaxagoras and Xenocrates,[5] and still more so that he rejected it because of the ignoble use made of it by the Stoics in favour of their 'rational god';[6] but there can be no doubt that he was familiar with Plato's *Timaeus*, and its claim that this world owed its origin to a mixture of reason and necessity.[7] I venture to submit, therefore, that Faustus, by rejecting the cosmogonies of Greek political

[1] W. Bauer, *Handb. z. N.T.*, 3rd ed., 1933, vol. VI.

[2] R. Bultmann, *Das Evang. des Johannes*, 10th ed., 1941.

[3] C. H. Dodd, *The interpretation of the Fourth Gospel*, 1955, 263 f., cf. 269 f.

[4] Cf. e.g. Plato, *Symp.*, 183A, *ἀρχὴν ἄρξαι ἤ τιν' ἄλλην δύναμιν*.

[5] Anaxagoras, cf. Aëtius, *Plac.* I.3.5; 7.5, ed. H. Diels, *Doxogr.*, 279, 299. Xenocrates, Aëtius, op. cit., I.7.30, H. Diels, 304.

[6] Cf. e.g. Aëtius I.7.33, M. Diels, 305, *νοερὸς θεός*.

[7] Plato, *Tim.*, 48A, cf. *Phaedo*, 97C etc.—On Goethe reading *Timaeus* cf. H. Trevelyan, op. cit., 251 n.2.

and rational philosophy, in his third endeavour chose the path of Greek mysticism.

The Hellenistic character of John I has not always found so little attention as it is finding today. In the early days of the Christian Church it was a view, held especially by the more learned Fathers such as Clement of Alexandria and Hippolytus of Rome, that the prologue of the Fourth Gospel was in essence identical with that of Heraclitus' book *On Nature*, a book which had earned him the title of the dark philosopher of Ephesus. Modern classical scholarship has discredited this report;[1] and it is regrettable that just those theologians whose concern with the Hellenistic elements in early Christianity is well known, W. Bauer and R. Bultmann, have divested themselves somewhat too quickly of this piece of evidence.[2] The way is thus opened for the claim that the 'pre-existent Logos', a conception of which it is said that it had its roots in Near-Eastern, and perhaps more particularly in Jewish mysticism, penetrated here into Christian thought.[3] If all this should hold even a spark of truth, and such I believe is indeed the case, then Goethe's reluctance to accept John I:I as it stands becomes understandable. Words, even 'the word', leading an existence of their own, were alien to the Hellenic mind;[4] and the Stoic concept of the Logos, found apparently in John I:I by the Fathers just quoted, admittedly had a classical ancestry; but at the time when the Fourth Gospel was written, it had to serve as an adaptation for more than one non-classical thought.

It may be held that the formula 'in the beginning', ἐν ἀρχῇ, was even

[1] Cf. W. Kranz's remarks in Diels–Kranz, 5th ed., I, 150, where references are given. The express statement is found in Amelios, the Neo-Platonist, quoted by Eusebius, *Praep. Ev.* XI.19.1, ed. Mras, II, 1956, 45, cf. E. Norden, *Agnostos Theos*, 2nd ed., 1929, 348 f.

[2] Even when accepting W. Kranz's verdict, which seems reasonable, it remains to be asked (*a*) who invented the ascription, and (*b*) how did it commend itself so widely? E. Norden, op. cit., gives the only answer I have found, and that a somewhat unsatisfactory one. A more thorough enquiry might well throw an unexpected light upon the relations between the early Fathers and late Stoicism, not to be found in the otherwise useful book by M. Spanneut, *Le Stoïcisme des pères de l'église*, 1957.

[3] Cf. G. Quispel in *The Jung Codex*, ed. F. L. Cross, 1955, 76, but note the warning issued by P. Billerbeck, II, 303, 'in fact, it cannot be held at all that the term of *memrah* was used consistently in the *targumim*'. The more positive remarks by G. G. Scholem, *Major trends in Jewish mysticism*, 1955, 114 f., refer only to the post-Christian era.

[4] C. H. Dodd, op. cit., 264 n.2, refers to Homer's 'winged words', but their wings only served the flight from mouth to ear.

B

nearer to Jewish, and more alien to Greek thought than the idea of the Logos. In Greek literature these words do not occur very often in a theosophical or philosophical meaning, whereas *maaseh bereshith* 'has always formed one of the main preoccupations of Kabbalism'.[1] However, such an argument from silence is not a very safe one when the fragmentary state is considered in which the Hellenistic theological and cosmological literature has come down to us. There is at any rate one saying that might at least be described as a near miss. This occurs in the short pamphlet *De incredibilibus*, written in the second century A.D. by a certain Palaiphatus. In the prologue to this treatise, which attempts to give rationalist explanations for various Greek myths, the author says:[2] 'For ever will I praise the companions Melissus and Lamiscus of Samos, who say: "In the beginning exists that which has come into being, and will be now".' The meaning given to this saying in the context is rather superficial. It is quoted to support the view that one should not assume that something which does not happen now might have happened in the distant past. However, since the saying is ascribed to two Pre-Socratics, an Eleatic and a Pythagorean, it may be assumed that originally it had a deeper significance than the Hellenistic rhetor, who has preserved it for us, may have realized.

There are, however, two preliminary problems to be attended to, before the meaning of the quotation can be discussed: its exact wording, and its origin. H. Diels had passed the saying, as it is quoted here, in his last edition (the fourth) of the fragments of the Pre-Socratics.[3] It was W. Kranz, revising Diels's collection for its fifth edition, who claimed that the words ἐν ἀρχῇ belonged to Palaiphatus, not to the quotation, and that the quotation as such was spurious.[4] Now it is obvious that if Kranz's division of the fragment were correct, it would no longer have a claim to figure as a parallel to John 1:1; and if the fragment were spurious, it would be necessary to show its pre-Christian origin in order to prove its relevance for our examination. For Palaiphatus, as we have seen, was a post-Christian author. However, it can be confidently claimed that Kranz's restitution of Palaiphatus' quotation would never fit the context. If 'that exists which has come into being, and will be

[1] G. G. Scholem, op. cit., 20.

[2] Palaiphatus, ed. N. Festa = *Mythogr. Gr.* III, ii, 1902, p. 2, ἐν ἀρχῇ ἔστιν ἃ ἐγένετο καὶ νῦν ἔσται.

[3] G. Delling in *Theol. Wörterb.* I, 1933, 477, 34; 478, 24, dependent upon this edition, has not found occasion to voice any suspicion.

[4] Diels–Kranz, 5th ed., 1934, I, 276.

now', Kranz's restitution, is not absolutely meaningless, it does not at any rate convey the ontological urgency by which the 'demythologizing' of people like Palaiphatus and his fellow Stoics was prompted. And it was a very similar spirit which prompted Palaiphatus to disbelieve the ancient Greek myths, and Dr Faustus to translate the Logos as 'the deed'.

Palaiphatus, like Faustus, 'cannot accord such high rank to the word'. He was an opponent of the 'nominalism' in the interpretation of ancient myths which began to make itself felt in the schools of his time, and continued till the early middle ages.[1] He therefore stated: 'It appears to me that all things that are named (πάντα τὰ λεγόμενα) have first come into being. For they cannot have arisen as names only; for in that case no word would have been coined for them. But the deed was done first, and then there was found a word for it.' Here, therefore, in a Greek author of the second century A.D., we find Dr Faustus' thesis that not the word was 'in the beginning', but the deed; and his treatment of the question suggests plainly that Palaiphatus intended making a general statement about coming into being, which was to embrace not only the various beginnings within this world, but—to use an as yet unexplained and therefore unguarded expression—the original 'beginning' of all that takes place in the world.[2] For Palaiphatus has made it plain that his thesis 'the deed was done first', refers to a creative activity 'in the beginning', which means that the words ἐν ἀρχῇ were contained in his quotation from Melissus. Without these two words the Pre-Socratic would not have lent any support to Palaiphatus' 'realistic' thesis. It therefore seems unwarranted—not to say ludicrous—to suggest that these words were

[1] What is meant here may be illustrated from the prologue to the second Vatican mythographer, ed. Bode, 1834, 74, *ab actibus autem vocantur ut Mercurius, qui mercibus praeest*. This sentence was taken from Isidorus of Seville, *Etym*. VIII.11.3, who in turn had culled it from a Latin grammarian of the early fourth century. He had claimed that the name Mercury was derived from Latin *merces*, and presumably ascribed to the most excellent merchant among the ancient heroes, according to the doctrines of Euhemerus. Here, therefore, the name existed before the god.

[2] Our criticism of W. Kranz's interpretation of Palaiphatus is concerned with fundamental questions. In his index to the *Vor-Sokratiker*, Kranz has tried to distinguish between ἀρχή meaning 'the beginning', and meaning 'a beginning'. Aristotle, at the latest, has given the proof that such an attempt is logically impossible; and when M. Luther added to his explanation of the first article of the Apostles' Creed, 'I believe that God has created me etc.', in his Shorter Catechism the words 'and still supports me', he may well have hoped to make this recognition popular amongst his 'dear Germans'.

meant to indicate the place where to find the quotation,[1] whilst it is highly improbable, not to say impossible, that Melissus' book in its integrity could still be found at all. Palaiphatus wanted his readers to know that Melissus had written: *ἐν ἀρχῇ ἔστιν ἃ ἐγένετο καὶ νῦν ἔσται*, 'in the beginning exists that which has come into being, and shall be now'. All that the Hellenistic author has done is to make us aware of a certain polemical edge, which is contained in the maxim of the Eleatic philosopher. All that seems to arise in this world, he claimed, is contained in it *in nuce*, nothing can be added, nothing lost bar—but Palaiphatus does not care for them—bar words.

We insist then that Palaiphatus quoted an anthology, and the question to be answered is simply whether the doxographer, the compiler of this anthology, gave his quotation a correct ascription. He chose two names, Melissus and Lamiscus. Melissus is by no means a mythical figure. He is known to have commanded a Samian fleet, which in the Samian war of 440 B.C. gained an initial success over an Athenian squadron, aboard which had fought Sophocles, the Athenian tragedian.[2] Our fragment, therefore, does not belong to those numerous ones that were ascribed to the somewhat obscure members of the early Pythagorean school who became the worshipped idols of Neo-Pythagoreanism. Quite apart from two larger treatises ascribed to Timaeus Locrus and Ocellus Lucanus, there are over a hundred of such fragments stored in the collections of John Stobaeus, not one of which can be judged genuine. It was largely for this reason that W. Kranz adjudged our quotation spurious. For whilst Melissus is well known, Lamiscus belongs to that group of doubtful Pythagoreans. Mentioned once (if we neglect the post-classical evidence) in Plato's seventh Epistle, as a pupil of Archytas, he appears as Plato's younger contemporary, more than half a century younger than Melissus. Whether this was the reason why he was described as a Samian in Palaiphatus' source, thus being duplicated and added as yet another to those obscure Pythagorean authorities,[3] remains an open question.

[1] An attentive perusal of H. Diels, *Doxographi*, will make it evident even to the very credulous that in the second century A.D. the works of the Pre-Socratics were no longer available to the general reader, and that only a charlatan would pretend to have consulted the original. We now have an excellent survey of the technical problems facing the preservation of the works of Roman lawyers by F. Wieacker, *Textstufen klassischer Juristen*, 1960, cf. esp. p. 142 f.

[2] Cf. E. Meyer, *Gesch. d. Altertums*, 2nd ed., 1912, IV, 65 f.

[3] The sources for Lamiscus, pupil of Archytas, who does not even figure in Überweg-Prächter, *Philos. d. Altertums*, 12th ed., 1926, are Plato, *Epist.* VII, 350B, the spurious answer to it in Diog. Laert. VIII.4.4. 80, and our passage from Palai-

It would be tempting to assume that the combination Melissus–Lamiscus arose from a confusion of two consecutive entries in the anthology used by Palaiphatus, the first ascribed to Melissus, the second to Lamiscus. The only trouble in this is that Lamiscus has not been credited by any ancient authority with a writing of his own. Thus a fairly strong suspicion that our saying should not be regarded as belonging to Melissus is justified. However, it seems that W. Kranz, referring for comparison to Melissus frg. B 1, has spoilt the case for spuriousness. This fragment reads:

What has been has always been, and will always be. For if it had come into being, before its coming into being it would without fail have been nothing. If therefore it was nothing, it could not in any way have become something out of nothing.

There is no contradiction here to our saying, 'in the beginning is all that has come to be, and that is going to be now'. I rather feel that this statement is needed to explain the 'pre-existence of being', which is contained in the Eleatic doctrine which is proclaimed in frg. B 1. This fragment is in fact incomplete, unless such a meaningful term as the *ἐν ἀρχῇ* of Palaiphatus is added as an explanation. The evidence from sensual perception for the coming into being of e.g. the leaves in spring is so strong that without such an explanatory addition frg. B 1 might appear as sheer nonsense. Therefore, unless we join Melissus to the various Sophistic clowns who made havoc of Parmenides' teaching, and for this there is no historic evidence, Palaiphatus' sentence cannot well be missed.

A word has yet to be said about 'Lamiscus the Samian'. If the pupil of Archytas was meant by Palaiphatus, it is unimaginable how he should have co-operated with Melissus in the production of a cosmology. Relations between Eleatics and Pythagoreans were not exactly cordial during the fifth century B.C., even if Empedocles, the philosopher who most strongly opposed Parmenides' teaching, was at best a rather heretical follower of Pythagoras. Palaiphatus, when he described Lamiscus as the 'co-author' (*συγγραφεύς*) of Melissus, and already the anthology which he despoiled, had a different mental image of the 'classical' period of Greek philosophy. Popular scholarship in the Hellenistic period had forgotten the tensions and jealousies of old. The romanticism of the Neo-Pythagorean forgers would not hesitate in the least to turn Lamiscus,

---

phatus. Nowhere is he described as a native of Tarentum. It is therefore likely that the 'Lamiscus of Tarent' in the index to Diels–Kranz, 5th ed., III, 533, is W. Kranz's own invention.

the pupil of Archytas, into Lamiscus the Samian. Neither can it be maintained that the basic philosophical differences between the two schools could have created any great obstacles for the new companionship of Melissus and Lamiscus. The history of Pre-Socratic, and generally of early Greek philosophy was little known amongst the authors of the Roman period. The first century B.C. saw Vergil ascribing a cosmological theory of Epicurean provenance to Orpheus;[1] and the production of yet another 'father' of Neo-Pythagoreanism would have had nothing uncommon. The character of these and similar inventions was mystic-romantic, which means that the secrecy of their doctrines so far would make them only more valuable. The magician and astrologer Nigidius Figulus, Cicero's friend, was a leading spirit in these Neo-Pythagorean circles, where Dr Faustus would have been welcome. And here lies the real importance of our parallel: Faustus with his translation 'in the beginning was the deed', took a decision in favour of a mystic doctrine of creation rather than of a scientific or philosophical one.

'In the beginning', these words introducing the Fourth Gospel, these words used apparently without very great emotion by the old Samian admiral Melissus in his contemplations 'On nature and being', took the fancy of these Neo-Pythagoreans. They conjured up a new father of their school, in order to support the doctrine of creative action as much by authority as by argument; and they handed to him a ready-made saying, which would otherwise not have had any legitimation in these circles. Having thus explained the twilight into which a genuine and serious philosophical thesis can be brought by somewhat unscrupulous mystics, we may now compare it with a proper forgery of the Neo-Pythagorean school: the spurious fragment ascribed to the famous early Pythagorean Philolaus, on the origin of the world:[2]

[1] Vergil, *Ecl.* VI.30 f.,

Nec tantum Rhodope miratur et Ismarus Orphea.
Nam canebat uti magnum per inane coacta
Semina terrarumque animaeque marisque fuissent
Et liquidi simul ignis: ut his ex omnia primis,
Omnia, et ipse tener mundi concreverit orbis;
35 *Tum durare solum et discludere Nerea ponto*
Coeperit et rerum paulatim sumere formas etc.

The Epicurean ancestry of these verses is attested by Servius in his commentary, ed. Thilo III, 1887, 69 f.

[2] Diels–Kranz, 5th ed., I, 417 f., cf. J. Moreau, *L'âme du monde*, 1939, 145 f.

Therefore the cosmos also continues, imperishable and unconquerable, through limitless eternity. For there is neither to be found an intrinsic reason which is mightier than she (W. Kranz: the cosmic soul), nor an external one that could destroy it. Rather is it a fact that this cosmos has existed from eternity, and will continue eternally, one only, and directed by the only one who is of its kindred (*συγγενής*), the mightiest, the unsurpassed. But the cosmos also contains the principle (*τὰν ἀρχάν*) of motion and of change, since it is one and continuous, and inspired by genetic power (*φύσει διαπνεόμενος*) and is from everlasting (*ἐξ ἀρχιδίου*). And its one part is immutable, but the other mutable. And the immutable extends from the cosmic soul, enveloping the universe, to the moon; but the mutable (*μεταβάλλον*) from the moon to the earth.[1] But since both the moving [force] causes gyration from eternity to eternity and the moved [substance] is conditioned according to the way the moving [force] guides it, it is necessary that the one is eternally causing motion, and the other is eternally passive. And the one is the universe, the turning[2] of the Nous and the soul, the other that of coming into being and change (*μεταβολή*). And the one is the first principle by its power and domination; the other the second and dominated. The sum of these two, however, the ever speeding divine,[3] and the ever changing creation, is the cosmos.

In this elaborate account of Neo-Pythagorean cosmology an attempt has been made to harmonize the two ideas of the eternity of the cosmos, and of its 'beginning'. The word *ἀρχίδιον*, if it can be passed at all, is evidently meant as a conflation of *ἀρχή*, 'beginning', and *ἀίδιος*, 'eternal'. This dilemma between eternity and 'beginning', I believe, is inescapable once people think in categories of time and causality; and it

[1] It is this widespread conception which has an analogy already in the pair *σαλευόμενος–ἀσάλευτα*, Heb. 12:27/28. If this is accepted, Windisch's, *Der Hebräer-Brief*, 2nd ed., 1931, 115, equation of *μετάθεσις* with *ἀπώλεια* is incorrect, although there is some patristic support for it, whereas Jerome says, *translationem*. W.'s conclusion that 'the eschatology of Heb. knows of no renewal of the created world', is therefore unsafe. For it seems far more probable that the *μετάθεσις* of Heb. means the same as the *μεταβολή* of Ps. Philolaus. It should be realized that the fragment of Apc. Petri preserved in Macarius Magnes IV.6, E. Klostermann, *Apocr. I* (Kl. Texte 3), 2nd ed., 1908, 13 no. 4, saying that heaven and earth will be judged, protests against the view expressed in Heb. 12:27/28.

[2] W. Kranz, op. cit., 418, puts a crux to *ἀνάκωμα*, and translates it 'Aufenthaltsort'. Yet Liddell-Scott show that the word, as well as the related *ἀνάκωσις* is found elsewhere also in the sense of 'the turning of the stars'. Ps. Philolaus, therefore, here seems to apply Plato's doctrine of the Nous and the soul as the self-moving movers, to be found, it is true, in the sphere of the fixed stars 'enveloping' the sub-lunar sphere.

[3] The play on words *τὸ ἀεὶ θέον θεῖον*, caused by the underlying astral theology, shows Stoic influence.

was the chief subject for Greek philosophy from the days of Anaximander to the time of the New Testament. The marks of this discussion make themselves frequently felt. Heb. 12:27/28 might be quoted as particularly close to our passage from Ps. Philolaus. This author in his turn is representative for a school of thought, which was prominent in Stoicism, but had its roots already in Aristotle, and even in Anaxagoras, where the active principle in the cosmos, *τὸ θέον θεῖον*, the speeding deity, was most highly exalted. Ps. Philolaus calls it 'first', a name which would be meaningless if the idea of 'beginning' were not taken into account. For this Neo-Pythagorean, as for Dr Faustus, it can be stated with conviction that he claimed 'in the beginning was the deed'.

It is of great significance that our footing amongst the contemporaries of St. John has been gained in the Neo-Pythagorean camp. Admittedly, Ps. Philolaus was dependent upon Zeno, the founder of Stoicism, who had held: 'The beginnings of the universe are two, the active and the passive principle', and even: 'God and qualityless matter';[1] but the Neo-Pythagoreans with their 'ever speeding deity' were the mystics of Hellenism, relying on vision and intuition rather than on logical deduction. Logically there are numerous holes in Ps. Philolaus' argument, some of them of a venerable age; but emotionally he receives much support from the quarters where he has his spiritual home. The Neo-Pythagorean ideas are clearly alluded to in an Orphic poem, called 'Diathekai', the mystic deposits or 'testaments' of the seer, which was frequently quoted by Jewish—and Christian—apologists. The Neo-Pythagorean ideas are most prominently displayed in those lines which have to be ascribed to a Jewish—or more likely a Christian—forger of the early second century, even if there are reasons why the original poem may have been of a pre-Christian, 'Pythagorean' origin.[2] The verses run as follows:[3]

[1] H. v. Arnim, *S.V.F.* I, 24, 5 f.

[2] Mystical cosmogonical poetry has a long history in Greek poetry, going back to Hesiod's 'Theogonia'. How influential it was in the circles of the educated at the beginning of the Christian era may be seen from Diodor. I.7, and the use made there of a quotation of Euripides' *Melanippe*.

[3] *Orph. frg.* 247, 22 f., ed. O. Kern, *Orph. frg.*, 1922, 261, taken from Eusebius *Praep. Ev.* XIII.12.5. Eusebius there claims the notorious Jewish forger Aristobulus as his authority; but it is unlikely that the forger was ignorant of John 1:18, 'no one has seen God ever; but God's only Son . . . has made him known'.—On the other hand, the claim by A. Boulanger, *Orphée*, 1925, 81, that Aristobulus was the author of the *Diathekai*, seems equally improbable. These forgeries have a way of growing in stages.

For none of mortal men can see the ruler,
Except the only begotten, the top of the tribe from on high
Of the Chaldees. For he knew the path of the star,
25 And of the sphere that for ever swings round the hub,
Cyclically, and equal in accordance with its axle.

There is no reason for us to worry just now about the spuriousness of these lines, although their connection with that Orphism which the classical tragedians and Plato had known is indeed slight. Neither need we go deeply now into the tortured question about the relations between Orphism and early Christianity.[1] We are moving in a company of warlocks and witches, and in such dissipated circles it even appears as satisfactory when we find the two religions in such close vicinity.[2] However, even this is only of remote concern. The fact which we wish to establish in this presumably Christian forgery is the identity of the supreme sphere in the Orpheo-Christian fragment with that uttermost sphere in Ps. Philolaus, the sphere which envelops the universe, and embodies the Nous and the cosmic soul. For this purpose it is necessary to realize that the sphere described in verse 26 of the Orphic fragment is the last stage on the cosmic journey of the Monogenes, the only begotten, through the heavens. In the same light, so it seems, the last sphere appeared to Ps. Philolaus: Beyond the 'ever speeding divine', and beyond the ever rotating ultimate sphere, there lay for the Neo-Pythagoreans the inscrutable mystery.

Having proceeded so far in relating the Orphic fragment to Neo-Pythagoreanism, we can now leave Ps. Philolaus. He has served us to mark the connection between the Johannine concept of the 'beginning' and the mysticism of contemporary non-Christian philosophical sects, but it is not claimed that the fragment quoted provided the immediate source for our Christian-Orphic fragment, let alone for the Johannine prologue. What I want to claim is the existence of a common theosophical-magical atmosphere in very early Christian and contemporary non-Christian circles. I want first of all to remind us of the fact that there was an apocryphal Christian tradition which held to the same Logos-Christology as the prologue of the Fourth Gospel, and had arisen at the time when this was written. This is found in the fragments of the

[1] Cf. W. K. C. Guthrie, *Orpheus and Greek religion*, 2nd ed., 1956, 261 f.

[2] Cf. A. Loisy, *Les mystères païens*, 1914, 49, 'la tradition particulière de l'orphisme ne donna point naissance à une secte organisée . . . elle aboutit spéculativement à des théories panthéistes et à un symbolisme subtil, pratiquement à une sorte de magie'.

Kerygma Petrou, which Clement of Alexandria quoted in his *Stromateis*:[1] 'Recognize then that there is one God, who made the beginning with all things, and has discretion over the end', such is 'Peter's' first thesis; and he continues with a description of God, calling him, 'the invisible who sees all, the immovable who moves all, the free from want whom the universe wants, and through whom it exists; unreachable, everlasting, indestructible, uncreated who created all things by the power of his word'.

If Dr Faustus had read this, perhaps he had not demurred; but now that we read it it is another observation that comes to the fore: The man who wrote these lines was not a Greek; he was certainly of Jewish origin. 'Recognize that there is one God': Wherever we find *εἷς θεός*, and have the least means of identifying the pen that scratched it, we find a Jewish hand that used the pen.[2] This certain result of research, largely in the field of ancient epigraphy, makes us listen still more carefully to our Orphic tradition. There are three different traditions of the 'Diathekai' and we have now to change over to one which was culled from yet another notorious forger by a Christian apologist. In the former, ascribed to Ps. Aristobulus by Eusebius, we found the 'only begotten' of John 1:18; in the following, ascribed to Ps. Hecataeus by Ps. Justin the author of the pamphlet *De monarchia* of the middle of the second century, we find the Word in the following exhortation:[3] 'Looking upon the divine word, attend to this [also]'.

In order to follow this Judaeo-Christian circle a little more closely to its lair, we have to return, however, to the Aristobulus fragment. As

[1] *Kerygma Petrou* frg. 2 = Clem. Alex., *Strom.* VI.5.39, ed. E. Klostermann, *Apocr. I* (Kl. Texte 3), 2nd ed., 1908, 13.

[2] That has been shown in that most careful study by Erik Peterson, *Heis Theos*, 1926, who, unfortunately, has not dealt with our fragment. He mentions, however, 241 n.1, the acclamation 'one Dionysus' in Orphic mysteries.

[3] The verse quoted, which is also to be found verbatim in the Aristobulus tradition, *Orph. frg.* 247, 6, seems to belong to the earliest lines of the Orphic 'Diathekai', which means either to a pagan setting, which would mean the Hermes Logos, or (which I think improbable) to a Jewish production. Clement of Alexandria, *Orph. frg.* 246, Kern, quoted the same line in yet another version of the poem. It is this line which makes me hesitate to accept the thesis of A. Boulanger, *Orphée*, 1925, 81, that the 'Diathekai' were a Jewish invention. Sources for a pre-Christian Jewish Logos-theology are so very rare!—The popularity of the Aristobulus tradition in Alexandrian circles may be seen from the fact that 'God's footprints', *Orph. frg.* 247, 19 Kern, are quoted in the Coptic *Gospel of Truth* 37:25.—On the use of existing literary material by Ps. Hecataeus cf. H. Willrich, *Urkunden-Fälschung in der hellenist.-jüd. Lit.*, 1924, 13.

we have said before, the Ps. Hecataeus fragment is another, and perhaps somewhat earlier form of the same Orphic poem, and shorter. The lines now to be quoted are in that part of the poem which has only survived in Eusebius–Aristobulus. This is how they run:[1]

> . . . He is everywhere,
> He himself, who is over the heavens, and achieves all things on earth,
> 35 Holding himself the beginning, middle, and ending,
> As the word of the Elders, as the son of matter ordained it.

Once more we are met with a conflation of Orphic, Jewish, and Christian formulae: 'Holding the beginning, middle, and ending', is a line with which already Plato had been familiar,[2] and which in the surroundings where we find ourselves at the moment, is very close to that Johannine text, Rev. 1:18, 'I am the first and the last and the living one', as well as to its Old Testament parallels, particularly Isa. 41:4. If we then add from the parallels in the Sibylline oracles those which proclaim the 'one God', which O. Kern has collected, and the remark contained in a scholion to the 'Tübingen Theosophy', by which it is shown that the 'son of matter' is Moses, 'who described the creation of the material world', and finally from the *Letter to Diognetus* the remark about the Logos developing into the Son, in the appendix, IX.4–5, which in its turn seems to play a part in the anti-monarchian struggles of Hippolytus of Rome, it will be felt that he who chooses the magician for his guide must not complain when he is landed in the witches' cauldron.[3]

I do not believe that this my last remark has been unjust to Goethe. He, the erstwhile acolyte of the Frankfurt Moravians, and well acquainted with, and deeply interested in, the medieval descendants of ancient Pythagoreanism and Orphism, would not have missed the biblical analogies mentioned here, any more than that between verse 34, 'He himself who is over the heavens (ἐπουράνιος), and achieves all things on earth', and Eph. 1:20 f., Christ 'enthroned in the heavenly realms (ἐπουρανίοις), far above all rule and authority etc.' Many mystical

[1] *Orph.frg.* 247, 33 f., Kern.

[2] Cf. *Orph.frg.* 21, Kern, and note the remark by W. K. C. Guthrie, *In the Beginning*, 1957, 111 n.7.

[3] O. Kern, *Orph. frg.*, 1922, 258, quoting Orac. Sib. III. 11; ibid., *Prol.* 94. *frg.* 1, 7. For the Tübingen scholion cf. O. Kern's remark, op. cit., 263, ad v. 34; On the *Letter to Diognetus* cf. H. G. Meecham, *The Epistle to Diognetus*, 1949, 145 f. It may also be mentioned that Hermes (Logos!) figures as the creator in the so-called Strasbourg cosmogony, cf. R. Reitzenstein, *Zwei religionsgeschichtliche Fragen*, 1901, 53.

ideas, particularly that of the 'Logos spermatikos', the germinating power of the word, can also be shown to be present in these Jewish-Orphic lines.[1] In short, we see here a religious world in ferment. And I maintain with full conviction that one, perhaps even the most important mystic word to set it in ferment, was *ἀρχή*, the 'beginning'. It has not deserved the uninterrupted neglect which it has endured from the commentators of the Fourth Gospel, at least since the beginning of the nineteenth century.

[1] Cf. e.g. *Orph. frg.* 245, 16, Kern (Ps. Hecataeus), 'containing the seeds of all being within his own body', which is said of Zeus, in conformity with an ancient Orphic tradition, cf. W. K. C. Guthrie, *Orpheus and Greek religion*, 2nd ed., 1952, 74; but that—in the mirror of the witches' cauldron—is really a very small matter. For in these circles almost any god can take the place of any other god.

PART ONE

The Classical Philosophers from the Pre-Socratics to Aristotle

2 The origin of the Greek idea of the 'beginning'

Whoever has read the Fourth Gospel knows that its first words re-echo the first words of Genesis, 'in the beginning', but there may not be many who ask the question: The beginning of what? There can be no doubt that the author of Genesis meant the beginning of the world, or perhaps better, the beginning of God's creative activity; but, despite a certain amount of evidence to the contrary to which I have drawn the attention of scholars a few years ago,[1] it is generally held that the Greeks were convinced of the eternity of the cosmos, and little research has been made into the response that a Greek, to whom after all the Fourth Gospel was addressed, would make to its opening: 'In the beginning.' It may be mentioned in passing that St. John took the inspiration to make the conception of *ἀρχή* an integral part of his theology not only from Gen. 1:1, but also from St. Paul who, in Col. 1:16 f., has made a very weighty statement, anticipating much of what is to be found in the prologue of the Fourth Gospel; it may also be remarked that in a different way the author of the Epistle to the Hebrews, especially 7:3, has made a characteristic use of the term. It is nevertheless true to say that in a special way the word is characteristic of the Johannine writings so that our starting point has to be: St. John and the Greeks.

To justify this starting point it is necessary to anticipate the result of an enquiry which for reasons of temporal sequence will have to follow in a later chapter. It can and will be shown that not only St. John's approach to the problem of 'the beginning', but also that of his Jewish contemporary, Philo of Alexandria, is essentially Greek in spite of the fact that they both had in mind the first words of Gen. 1 in their Septuagint form.[2]

[1] A. Ehrhardt, 'Creatio ex Nihilo', *Studia Theologica* IV, 1950, 13 f.

[2] Cf. *infra*, pp. 193, 196.—Josephus, *Ant.* I.1.1, begins his Jewish Antiquities with a reference to Gen. 1:1, but side-tracks the problem by saying: 'And this is the first day, which Moses calls one. The reason for this I could easily give even now. However, since I have planned to explain the philosophy of the universe (*τὴν αἰτιολογίαν ἁπάντων*) in a separate treatise I will defer my explanation of it till then,' but in *Adv. Apion.* II.22, where the matter is taken up again, no such explanation is attempted.

Here a short remark why this should be so, and was in fact inevitable, seems to be called for. It appears that the problem of 'the beginning' is in itself an ontological problem. It arises only on the presupposition that actually, i.e. as the basis for human action, the world has a continuous being 'in time'. For it is time which engenders the conceptions of 'beginning' and 'ending', whilst equally time makes it impossible to think of an absolute beginning 'in time'. For infinity of time—unlike the infinity of matter, which the Greeks came to describe, since Aristotle, as a passive infinity—is an active one which does not allow of acting upon it, but only of acting within it. This very recognition lends that haunting profundity to Luther's well-known reply to the question what God was doing before He created the world: 'He sat under a birch tree cutting rods for those who ask nosey questions.' The same recognition gives its significance to Philo's remark about the 'time' when the world was not yet in existence,[1] *ἦν ὅτε οὐκ ἦν*, for the evolution of the Christian doctrine about the Saviour and His work of Salvation, which came dramatically to the fore in the Arian controversy of the fourth century.

All this, however, is wholly dependent upon the acceptance of the ontological principle in the Greek world-view, which is largely, if not altogether, coincidental with the fundamentals of the cosmological theories of nineteenth-century philosophy. If, however, the world was not seen in this particular light the conclusions just outlined became meaningless. This fact was recognized in post-exilic Judaism. An evident word of warning against the Greek point of view was, so it seems, voiced in Eccl. 3:11. When the Preacher said, 'and he gave *ha'olam* in their heart', a word correctly translated in the Greek as *τὸν αἰῶνα*, he wanted it to be contrasted with *'itto*, His (God's) time. In Jewish theology '*olam*, 'the world' is not eternal, but is that time which is co-equal with God's creation. It therefore cannot be regarded as the world's *ἀρχή*, a fact which is also stressed repeatedly by Philo.[2] Consequently it is only the effect of their ignorance of being God's creatures by which the

[1] Philo, *De decal.* 58, Cohn–Wendland, IV, 281 f., . . . *καὶ γὰρ γέγονε <κόσμος>: γένεσις δὲ φθορᾶς ἀρχή, κἂν προνοίᾳ τοῦ πεποιηκότος ἀθανατίζηται καὶ ἦν ποτε χρόνος ὅτε οὐκ ἦν. Θεὸν δὲ πρότερον οὐκ ὄντα καὶ ἀπό τινος χρόνου γενόμενον καὶ μὴ διαιωνίζοντα λέγειν οὐ θέμιτον*, cf. *De opif. mundi* 26; *De spec. leg.* 1.266, Cohn–Wendland, I, 8, 5 f.; V, 64, 16 f.—The evident connection with Alexandrian theology (Origen–Arius) does not seem to have been generally recognized. Neither F. Loofs, *Leitfaden d. Dogm. Gesch.*, 4th ed., 1906, 234 f., nor G. Bardy in Palanque–Bardy–Labriolle, *De la paix const.*, 1947, 72, refers to it.

[2] Philo, *De opif. mundi* 26; *De aetern. mundi* 52 f., cf. 71 f., Cohn–Wendland, Reiter, I, 85, f.; VI, 89, 1 f.; 95, 3 f.

Greeks were caused to adopt their erroneous cosmology. In addition to the protest against such an incorrect Greek theory, Eccl. 3:11 even offered an alternative. In the words with which the verse begins, 'he hath made every thing beautiful in its time', there is to be found that ancient Hebrew theological doctrine of creation which has found its classical expression in the assertion of Gen. 1:31, that God viewing His creation 'saw that it was very good', and in Job 38:7/8, that 'the morning stars sang together, and all the sons of God shouted for joy'. These statements indicate the view that creation is the continuing expression of God's eternal goodness, and viewed from this angle, the question of earthly time completely vanishes from the problem of 'beginning'. The Greek word *ἀρχή*, however, remains serviceable since its combined meanings of 'beginning' and 'rulership' find their proper place in the Divine property of goodness. Hellenistic Judaism as represented by Philo was still fully aware of the existence of this Hebrew alternative.[1] It was the problem of evil, so clearly put before the eyes of the Jews by the non-Jew Job, 'the man in the land of Uz', which was responsible for the change of this teleological world-view to an ontological one, even in the Jewish mind.[2] If, because of its Greek origin, the characterization as 'teleological' should be regarded—not without reason—as giving only an approximate definition of the Hebrew theology of creation, I would refer for its interpretation to the detailed account of it that has been given recently by T. Boman.[3]

It seems to me very significant that the change of mind which we have observed among the Jews with regard to God's creative activity appears to find an analogy in the development of Greek thought. There is reason to believe that in Greek cosmology too the introduction of the conception and the term of *ἀρχή* was connected with the discovery of the

[1] Philo, *Legg. alleg.* III.78, Cohn–Wendland, I, 130, 1 f., *τοῖς γοῦν ζητοῦσι τίς ἀρχὴ γενέσεως, ὀρθότατα ἂν τις ἀποκρίνοιτο ὅτι ἀγαθότης καὶ χάρις θεοῦ ἣν ἐχαρίσατο τῷ μετ' αὐτὸν γένει· δωρεὰ γὰρ καὶ εὐεργεσία καὶ χάρισμα θεοῦ τὰ πάντα ὅσα ἐν κόσμῳ καὶ αὐτὸς ὁ κόσμος ἐστί*, interpreting Gen. 6:8, 'Noah found grace in the sight of God'. Noah is also said by Philo, *De vita Mosis* II.60, Cohn–Wendland, IV, 214, 5 f., *δευτέρας γενέσεως ἀνθρώπων αὐτὸς ἀρχὴ γενέσθαι*, cf. 1 Pe. 3:18 f.

[2] W. Bousset–H. Gressmann, *Die Relig. des Judentums*, 3rd ed., 1926, 358 f., have most ably set out this change, and shown its connection with Job, pp. 385 f., which substituted justice and omnipotence for goodness and mercy as the pre-eminent ones amongst the divine properties (cf. the sources quoted on p. 359 n. 3) but have not laid much stress on its particular effect upon Jewish cosmology—but cf. their description of the cosmological significance of sin, ibid., 402 f.

[3] T. Boman, *Das hebräische Denken im Vergleich mit dem griechischen*, 2nd ed., 1954, 75 f.

problem of evil. Obviously, in Greek thought as elsewhere, the ontological approach to the phenomenon of the visible world was preceded by mythological and magical approaches. Both these approaches have left their marks upon the subsequent Greek philosophical cosmologies. It is also true to say that, once the term of *ἀρχή* was introduced, its value for the determination of the point where the infinite enters the finite, and the eternal the temporal was speedily recognized. That much may be seen from the fact that already Protagoras, the Athenian sophist, wrote a monograph entitled *Περὶ τῆς ἐν ἀρχῇ καταστάσεως*, although neither a summary of its argument nor even a single fragment from its content seems to have survived.[1] In later time the doxographers of the Hellenistic era used the word for a number of chapter headings in their anthologies of the opinions of the Greek natural philosophers. In this way they provide ample evidence for the fact that thoughtful men during the last three centuries of the pre-Christian era were greatly occupied with the question in what sense it might be held that this world surrounding man had a beginning. John 1:1, we may conclude safely, did not startle its readers by introducing an unconventional consideration.

This almost feverish interest taken in 'the beginning' by those Hellenistic collectors of ancient wisdom, who in the last instance always depended upon Aristotle and Theophrastus,[2] constitutes almost our only source of information on pre-Socratic thought about *ἀρχή*, and confronts us with a dilemma. For on the one hand Hippolytus of Rome reports, that it[3] was Anaximander who had introduced this term into Greek philosophical discussion, and on the other we learn from the fragments of the anthology of Aëtius that already Thales, the teacher of Anaximander, had held 'that the *ἀρχή* of things existing' was water.[4] Hippolytus' source was either Aëtius or his own imagination, and it seems somewhat difficult to ascribe to Aëtius such inconsistency. Nevertheless, this appears to be the only way out of the dilemma. Not only is it true to say that Hippolytus has been praised recently for his accuracy with regard to Anaximander,[5] but it seems certain that Aëtius' report on Thales is unacceptable. It may well have its origin in that spurious work *Περὶ ἀρχῶν*, ascribed to Thales by Galenus, which has long been

[1] H. Diels–Kranz, *Vorsokr.*, 5th ed., II, 255, 2, cf. Überweg–Prächter, *Philos d. Altert.*, 12th ed., 1926, 116.

[2] Cf. the enlightening remarks by Ch. H. Kahn, *Anaximander*, 1960, 12 f.

[3] Hippol., *Philos.* 1.6.2, cf. Diels–Kranz, 5th ed., I, 84, 3 f., frg. A 11.

[4] Aëtius, *Plac.* 1.3.1, H. Diels, *Doxogr.*, 276, 5 f.

[5] Cf. Ch. H. Kahn, op. cit., 15 f., who on p. 30 f. accepts and defends the interpretation of Hippolytus' saying as it is stated here.

recognized as a Hellenistic forgery.[1] The case for Anaximander, however, is different. We have to admit, it is true, that already in the Odyssey *ἐξ ἀρχῆς*, meaning 'from everlasting', does occur.[2] It is also true that in other respects Hippolytus cannot claim a convincing authority for his tales about early Greek philosophy.[3] However, even H. Diels, who otherwise regards the Church Father with the utmost suspicion, admits that a '*pretiosissimus opinionum thesaurus compilatur*' in his 'Philosophumena' with regard to Anaximander, i.e. that he collected in this instance his information from genuine references to Theophrastus' anthology. Our reliance upon Hippolytus' testimony is also supported by the fact that Anaximander had a reputation as an inventor and innovator in other respects as well.[4] For this fact makes us aware of the existence of a special interest in this part of his exploits, existing long before the time of Hippolytus; and it may only be an accident that this, his most momentous innovation, had not been mentioned by any contemporary or earlier source. It is, however, mentioned by the sixth-century Neo-Platonist Simplicius,[5] who may have had access to a copy of the original of Theophrastus' work.[6] In any case, Simplicius was not dependent upon Hippolytus. We may thus rely upon their reports, which are almost verbatim identical, as a conviction held already by Aristotle and his immediate pupils.[7]

If we proceed on this assumption that Anaximander really introduced the conception of *ἀρχή* into the philosophical discussion, the general opinion that he has to be grouped with the other 'natural philosophers' from Thales to Heraclitus,[8] who tried to establish a 'monistic' world-

[1] Cf. H. Diels–Kranz, 5th ed., I, 80 f., Thales frg. B 3.

[2] Homer, *Od.* I.188; II.254.

[3] Cf. the remarks about Hippolytus made by H. Diels, *Doxogr.*, 145, who calls him, and his sources for the doctrines of Heraclitus and Empedocles, '*indoctissimum scriptorem*'.

[4] Cf. Diels–Kranz, frgg. A 1, 2, 6, 7, where he is said again and again to have done something 'as the first'.

[5] H. Diels–Kranz, 5th ed., I, 83, 5, Anaximander frg. A 9.

[6] Since the publication of that excellent work on the effect of the change from scroll to codex, F. Wieacker, *Textstufen klassischer Juristen*, 1960, I have become still more hesitant to propose any such suggestion, and only repeat it as a view of Ch. H. Kahn's, because the library at Athens may have been in the possession of such a treasure even at that late date.

[7] In his review of Ch. H. Kahn's book, W. K. C. Guthrie, *Mind* LXX, 1961, 564, takes the historicity of the reports by Hippolytus and Simplicius for granted.

[8] Cf. Überweg–Prächter, *Philos. d. Altert.*, 12th ed., 1926, 41. Ch. H. Kahn, *Anaximander*, 1960, 199 f., who bases his views entirely on the fragment of Anaximander, may not have seen the whole extent of the problem of the relations

view, i.e. one that was founded upon one single 'element', stands in need of some considerable clarification. This is all the more true since they all, including Heraclitus, seem to have avoided the use of ἀρχή. The word nowhere occurs in the considerable number of the existing fragments of Heraclitus' writings in particular. It has, therefore, to be remembered that it was Theophrastus, the pupil of Aristotle, who aligned these Milesians—and others—under the point of view that they considered as ἀρχή the water, the air, the fire, the elements in general, or the atoms. Such a line of evolution of thought appears reasonable in retrospect; but even so it is necessary to admit that Anaximander's assertion that it was the infinite (ἄπειρον) which was the original 'beginning' cannot find any place in it. For this theory seems to approach the problem of the cosmos from without, whilst all the others clearly approach it from within. At the same time, assuming that Theophrastus was no idiot who overlooked this difficulty, his procedure is very significant. For by adopting it he shifted the emphasis from the predominantly spiritual character given to the ἄπειρον by his teacher, Aristotle, to a more materialistic conception of it. Anaximander's ἄπειρον, according to Theophrastus, consisted of four components, none of them material in the common sense, yet in their totality forming a system which was meant to contain all the approaches by which human reason could meet infinity: Infinite time, space, durability, and numbers.[1] At the same time the idea of matter, which probably was not yet clearly grasped by the Milesians, was evidently also very close to this system: On the one hand we find it described as a σῶμα ἄπειρον, and in early Greek thought the idea of an infinite body in the abstract, without any substance filling it, would have been meaningless. On the other hand, we hear that the system was described as μεῖγμα, a term which presupposes material constituents.[2] From this term it may be understood why Theophrastus

between him and the Milesian school. Abel Rey, *La jeunesse de la science grecque*, 1933, 56, is right in describing Anaximander's Περὶ φύσεως as 'une cosmologie et une physique largement débarrassées, dans le détail du moins, d'idées religieuses ou mystiques', if the parenthesis is given sufficient weight; but his assertions, p. 84 f., on the meaning of ἀρχή, and his belief that Aristotle still had access to Anaximander's complete work, whilst Theophrastus could get hold of only one single quotation from it, are unconvincing.

[1] Aristotle, *Phys.* III.4, 203b, 6 f., Diels–Kranz, 5th ed., I, 85, 14 f.—A similar shift of emphasis as that of Theophrastus, which amounts to misrepresentation, may be observed in Alexander of Aphrodisias, Anaximander frg. B 16, cf. H. Diels's remark, ibid., p. 85, 32 f.

[2] Cf. W. Spoerri, *Späthellenistische Berichte etc.*, 1959, 11, with an account of recent discussions, ibid., n. 3.

records it as Anaximander's view that matter, moving of its own, automatically, produced accidentally not only this, but also innumerable other worlds.

Though in this way the differences between the accounts of Anaximander's doctrine by either Aristotle or Theophrastus have been somewhat lessened, yet they cannot be entirely reconciled.[1] In this state of uncertainty scholars will differ already when translating the only substantial fragment of Anaximander's work that has been reconstituted with some degree of probability of resembling the philosopher's own words. Even when reconstituting it one important decision has to be taken, whether it is to be regarded as one fragment or as two.[2] I follow Diels–Kranz in treating the fragment as one, mainly because I believe that Theophrastus himself quoted from an earlier quotation; but I will mark the possible division by a paragraph:

> The ἀρχή of all that exists is the infinite.
>
> And whence is the coming into being for all that exists, thereinto also their destruction takes place as it is fitting. For to each other they pay fine and compensation for their unrighteousness, according to the ordinance of time.

In recent times Erik Wolf, the great German lawyer, has interpreted this saying, which he regards as one of the foundations of European legal thought, from the point of view of modern existentialism, basing his argument upon lectures given by M. Heidegger in the winter of 1941–2.[3] This attempt is of a particular interest because of its legal appreciation of the fragment, which has caused me to translate δίκη as 'fine' rather than as 'amendment'.[4] The principles of Heidegger's interpretation, on the other hand, do not seem convincing.[5] In order to

[1] Ch. H. Kahn, *Anaximander*, 1960, 22 f., giving a selection of Aristotle's frequent inaccuracies, seems to follow Theophrastus, whereas Abel Rey (*supra* p. 21, n. 8) comes down on the side of Aristotle.

[2] Ch. H. Kahn, *Anaximander*, 1960, 29 f., cf. 35 f., makes it clear that Simplicius did not introduce the whole sentence about the ἀρχή as *verba ipsissima*, but only that one word. However, on pp. 166 f. he treats the sentence as being substantially part of the fragment. Whilst agreeing with his decision, I regret that he has given no reason for his taking it.

[3] Erik Wolf, *Griech. Rechtsdenken*, I, 1950, 220 f.

[4] So Ch. H. Kahn, *Anaximander*, 1960, 166.

[5] It is Heidegger's contribution that Wolf compares the Anaximander fragment with Goethe's

> 'Nach dem Gesetz, wonach du angetreten,
> So musst du sein, dir kannst du nicht entfliehn.'

(*Urworte. Orphisch* [*continued overleaf*]

understand Anaximander it seems necessary to me to place his saying in its entirety into kindred surroundings, and Heidegger and Erik Wolf have failed to do this.[1]

In order to see Anaximander in his proper setting we have shown previously that his saying does not belong to that series of natural philosophers who considered from within the problem of the *ἀρχή* of the cosmos. The *ἄπειρον*, however we may define it, is outside the cosmos as Anaximander sees it. It is for this reason that we suggest that Anaximander's thought is more closely akin to Empedocles' thought than to Heraclitus and his 'monistic' precursors,[2] whom we have found concerned with a different problem. There are in particular three reasons why a comparison between the fragment of Anaximander and the remnants of Empedoclean thought might prove fruitful. The first is a terminological one, based upon the fact that the surviving sayings of Heraclitus nowhere contain the term *ἄπειρον* whereas Empedocles described the original state of the universe as an *ἀπείρων σφαῖρος*.[3] This use of the term is also significant for Empedocles whose close connection with the Pythagorean school is, of course, well known. For amongst the Pythagoreans *ἄπειρος* was a household word. The use of the term by Empedocles is, therefore, by no means unexpected. However, this use seems to cast a special light upon Anaximander. For it appears that when introducing the idea of *ἀρχή* into Greek philosophical discussion, he used a term *ἄπειρον*, which was to be appropriated by the Pythagoreans. Consequently it may be held that it was he, Anaximander, who gave to the schismatic Pythagorean, Empedocles, the opportunity of using his theory of the world's origin by couching it in familiar terms.

Providing that this is meant seriously, and not as a stylistic ornamentation, Heidegger, who must have been aware that Goethe did not know the fragment, seems to imply that (*a*) Anaximander's statement has a strong Orphic flavour, and (*b*) that it belongs to moral philosophy even more than to natural philosophy. Both these positions will have to be proved, I feel, by stronger evidence than such a quotation.

[1] E. Wolf, op. cit., 220 f., has almost atomized the fragment, and then compared the contemporary use of the various words constituting it, which would have been somewhat more convincing if the language were more archaic; but in spite of Simplicius–Theophrastus' assertion of its 'poetic' character, the fragment does not contain any notable archaisms.

[2] Cf. how Plato in Empedocles frg. A 76, and Theophrastus in frg. A 86, Diels–Kranz, treated Empedocles and Heraclitus as opposites. Erik Wolf, op. cit., 305, is mistaken when trying to explain Empedocles by comparing him with Heraclitus, the near-Pythagorean with the passionate opponent of Pythagoreanism.

[3] Empedocles frg. B 28, 1, Diels–Kranz, 5th ed., 1, 324, 10.

The second reason why reference to Empedocles should be made in order to shed some light upon the obscure meaning of Anaximander's fragment is that ancient witnesses refer to the close connection between Anaximander and Empedocles. Admittedly, the story told us by Diogenes Laertius[1] that Empedocles copied Anaximander's way of life by dressing in a sacred robe, like a tragedian, is a piece of Hellenistic slander, made improbable already by the time gap between the two.[2] The report shows, nevertheless, that already in the pre-Christian Hellenistic era a close connection between the two philosophers was regarded as an established fact. Since, however, there had been no chance for them to meet in the body, we may conclude that this connection was found in their respective philosophies. This conclusion is supported by the repeated references to their similarities and divergences in Aristotle's 'Physics'[3] which, viewed under this aspect, may assume a special significance. Thirdly and finally, it can be shown by a comparison between the fragment of Anaximander and the philosophy of Empedocles, better than in any other way, that the fragment has indeed to be treated as one coherent unit, and not as two separate sayings.

It has to be admitted that the first sentence of Anaximander's fragment constitutes a paradox; and it may well be asked in what way his statement, 'the beginning of all that exists' (which I understand as 'all that is finite'[4]) 'is the infinite', could be possibly resolved by the second sentence. It is for this very reason, I believe, that Ch. H. Kahn, on account of a purely grammatical observation made by H. Cherniss, stresses that the second sentence 'on merely grammatical grounds' contains no reference to the *ἄπειρον*.[5] In Ch. H. Kahn's subsequent considerations the first sentence

[1] Diog. Laert. VIII.70 = Anaximander frg. A 8, Diels–Kranz, 5th ed. I, 82, 37 f., quoting an otherwise unknown Diodorus of Ephesus.

[2] Whatever confidence we may put in Apollodorus' report, Anaximander frg. A I, Diels–Kranz, 5th ed., I, 81, 8 f., that Anaximander died shortly after 547/6 B.C., it is impossible to assume that he was still alive at the outbreak of the Ionian revolt in 500 B.C., the time when Empedocles was born.—On the other hand, it seems possible that Anaximander rather than Pythagoras was the anonymous person referred to by Empedocles frg. B 129, Diels–Kranz, 5th ed., I, 364, I f.

[3] Aristotle, *Phys.* I.4, 187a, 20; III.4, 203b, 6, Diels–Kranz, 5th ed., I, 83, 14; 85, 17 f.

[4] Ch. H. Kahn, *Anaximander*, 1960, 174, renders *τὰ ὄντα* as 'the things that are'. There is no objection to this on grammatical grounds; but when Simplicius, whom Kahn quotes on p. 166, emphasizes that the *ἄπειρον* is essentially different in Anaximander's view from the cosmos constituted by the elements. it seems necessary to express this difference in the translation.

[5] Op. cit., 167 f., esp. n.I. Cherniss's observation refers specifically only to Simplicius to whom the indirect speech of the quotation has to be attributed. Whether

is neglected so that the conclusion is at hand that he also despaired of finding a logical connection between the two sentences. In contrast to these views, I venture to suggest that the solution to the apparent paradox: 'The infinite is the principle of all existing finite things', can be derived from the three new propositions put by Anaximander in the second half of the fragment, necessity (*τὸ χρεών*), guilt (*ἀδικία*), and the order of time (*τάξις χρόνου*), not absolute time which, according to the doxographers, Anaximander rather regarded as a component of the infinite.

Of these three propositions the second and the third seem to have their evident equivalent in the philosophy of Empedocles. We find 'the order of time' in his *περιπλόμενος χρόνος*,[1] and the 'guilt' in his conception of *κακότης*.[2] As to 'necessity' it may be held that Empedocles by personifying *'Ανάγκη*, as he does, has somewhat obscured the analogy to Anaximander's less personal approach, describing it simply as *τὸ χρεών*. Of these three the most enlightening analogy between Empedocles' and Anaximander's thought is to be found in the conception of 'guilt'. First of all, there can be no doubt that Empedocles believed that it had been his own *κακότης* which had brought the result that he, who was essentially an eternal being, a god, had been constrained into the circumscribed, finite shape of a mortal human. It can also be claimed that *τὰ ὄντα* has in frequent cases the precise meaning of 'living beings'[3]; and that Erik Wolf and Heidegger in their 'existentialist' interpretation of Anaximander's fragment have concentrated, with good reason I believe,[4] upon this meaning. If furthermore it is admitted, as my friend

the plural used by Simplicius, or the singular used by Aëtius (p. 168), is genuine cannot, I believe, be decided by abusing Aëtius as 'banal and inaccurate'. He is also over 400 years nearer to the original. I hesitate to accept the principle proposed on p. 34 f., that Hippolytus + Simplicius = Theophrastus.

[1] Empedocles frg. B 17, 29, Diels–Kranz, 5th ed., I, 317, 10.—The 'order of time' here has nothing to do with Solon's Dike of Kronos, to which Ch. H. Kahn, op. cit., 170 n.4, refers.

[2] Empedocles frgg. 144, 145, Diels–Kranz, 5th ed., I, 369.

[3] For evidence on this cf. G. Zuntz, *Hermes* 87, 1959, 438.

[4] Ch. H. Kahn's theory, *Anaximander*, 1960, 178 f., regarding the conflicting beings in Anaximander's fragment as the elements, and finding no cogent reason for rejecting the assertion made by Simplicius that this was so, is unconvincing. Perhaps, the general distrust of the reliability of contemporary Church Fathers need not be extended to pagan authors (though I doubt it), but I hesitate to take the further step of admitting that the view proposed was obviously not Simplicius' notion of 'element', but that of Anaximander. For Simplicius, if he meant that, acted objectionably by suppressing the difference, as he also did if he introduced a concept of 'element' which was unknown to Anaximander.

Dr G. Zuntz assures me it should be, that Empedocles' 'Katharmoi' are to be used for the interpretation of his 'Physica' it may be held with some confidence, that Empedocles was not the first who taught that it was individualization by defilement which plunged the living (*τὰ ὄντα*) into being. Already Anaximander, the first among the Greek philosophers who enquired about 'the beginning', put this question because of a consciousness of guilt and defilement as well as because of its close connection with 'the end' and man's fear of it. Thus the process which was witnessed in Judaism by Ecclesiastes, of the change from a teleological to an ontological understanding of the cosmos, had its analogy in Greek thought, if at a slightly earlier date. In fact, when Ecclesiastes claimed that God fixed 'the aeon' in man's heart he meant roughly the same as Empedocles—and Anaximander—by subjecting the world and man to 'the order of time'.[1]

There may yet be a profound difference between the three, and especially between Anaximander and Empedocles, with regard to what they understood by 'necessity'. It appears to me a tempting hypothesis that Anaximander's expression *κατὰ τὸ χρεών* contains a reference not only to death, but even to the judgment on the dead, especially since it appears in such close proximity to Dike, a goddess who was closely connected in Greek thought with the last judgment.[2] However, this assumption is doubtful: Empedocles, it is true, had to teach the judgment of the dead because it was demanded by his theory of the migration of souls,[3] but he nowhere used the term *τὸ χρεών*. The expression is found, however, and in close connection with Dike too, in Heraclitus.[4] Yet this philosopher rejected the theory of the migration of souls, as held by Pythagoras, in quite general terms, and never in his known fragments so much as hinted at the possibility of a last judgment. Since it appears possible, on the other hand, that he borrowed this 'poetical term' from Anaximander, its connection with a judgment on the dead in Anaximander's philosophy has to remain uncertain.

[1] It seems to me that the meaning of *τάξις χρόνου* has found its most profound interpretation in Eccl. 3.

[2] On Dike and the netherworld in Orphism and early Pythagoreanism cf. my *Politische Metaphysik*, I, 1959, 39 f. On Dike and the judgment on the dead cf. *Studi in memoria Emilio Albertario* II, 1950, 547 f.

[3] Cf. on the details L. Ruhl, 'De mortuorum iudicio', *Religionsgesch. Vers. und Vorarb.* II.2, 1903, 14.

[4] Heraclitus frg. B 80, Diels–Kranz, 5th ed., I, 169, 3 f. On the meaning of Dike in Heraclitus cf. Erik Wolf, *Griech. Rechtsdenken*, I, 1951, 239 f.

3 Orphics and Pythagoreans

Two important facts about ἀρχή have so far been established, the date when, and the reason for which, the term was introduced into Greek philosophical discussion. These two facts will now be used to determine the original meaning of the word, and the way which our enquiry will now have to take. As to the latter, a thorough re-appreciation of sixth-century Ionian philosophy seems to be demanded, but cannot, of course, be undertaken here. It has to be realized that when Thales is said to have held that water was the ἀρχή of the cosmos this is already an interpretation, and probably a faulty one. For both these terms made their appearance in philosophical discussion after his time. From this it follows that more serious attention has to be given to the reports about the new introduction of basic terms into the earliest Greek philosophical discussion. Since it has become clear that it was not before Anaximander that the term ἀρχή was introduced into Greek philosophical discussion, just as it is said of Pythagoras that it was he who gave the special philosophical meaning to the term of cosmos,[1] these two new introductions coming from two philosophers who may have been in personal contact,[2] seem to mark the differences between the various pre-Socratic systems more convincingly than the modern distinction between 'monistic' and 'dualistic' systems.[3] The easy way in which the doxographers—from Theophrastus to the writers of contemporary text-books—have used the term ἀρχή for the description of pre-Socratic theories is doubtlessly misleading, and in fact a complete misrepresentation of the thoughts of these philosophers. For in their time the term was a controversial one,[4] a fact which appears already when the survival of the precise tradition

[1] Überweg–Prächter, *Philos. d. Altert.*, 12th ed., 1926, 86.

[2] Pythagoras is said by Überweg–Prächter, op. cit., 61, to have been represented by ancient writers as a pupil of Anaximander, cf. Jamblichus, *Vita Pyth.* 11 fin. The text has not been quoted by Diels–Kranz, but there is nothing intrinsically impossible in this notice, and it seems significant that the Neo-Platonists at any rate found the two philosophers in close contact. They knew something about Anaximander.

[3] O. Gilbert, *Griech. Religionsphilosophie*, 1911, is wholly given to this somewhat facile distinction, and it is on account of this, as I believe, that he takes notice of the conception of ἀρχή no earlier than in connection with Aristotelian philosophy, p. 425.

[4] Cl. Baeumker, *Das Problem der Materie*, 1890, 12 f., has rightly chastised Aristotle for his irresponsible neutralizing of Anaximander's views.

about the man who introduced it is critically considered. It would therefore appear as an uncritical proceeding if the views of those pre-Socratics who in their existing fragments never made use of the word ἀρχή were employed in the elucidation of its meaning, even if the word is to be found in the summaries of their teachings produced by later writers.[1]

This leads us to the task of defining exactly what was meant by the word ἀρχή when Anaximander introduced it into the philosophical discussion of the sixth century B.C. The etymological derivation of the word, which incidentally seems to be rather uncertain,[2] does not, I think, offer any help. More important is the fact that Anaximander used the word in the singular. For in this way it may be claimed for him that he was a 'monist'. This, however, does not imply that he should be described as a monistic materialist, although it has to be admitted that he assumed that the ἄπειρον consisted of matter. If the introduction of such a term should serve some useful purpose Anaximander's approach to the problem of matter might even be described with Cl. Baeumker as 'hylozoism'.[3] However, to my mind such considerations completely miss the salient point in Anaximander's cosmological theory. For such 'living matter', which in its eternal movement produces and annihilates accidentally innumerable worlds, 'exists' only dialectically by a process of reasoning. The 'existing' infinite (as distinct from any vague infinity) has, according to Anaximander, a material component which is totally different from any observable, finite matter, i.e. in the sense of natural science no material component. To him existence and rational apperception seem to have been as mutually dependent as they were mutually exclusive. Without an existing finite the infinite could not 'exist' any more than the cosmos could 'exist' without an existing infinite. This mental process of calling the cosmos and the infinite into 'existence' was, however, no causative process. For 'existence' cannot be regarded as a change in the condition of vague infinity.

[1] By this token all the atomists have to be dismissed from our enquiry, although a number of Diels's 'A' fragments of both Leucippus and Democritus contain the word. They are, however, useful for the illustration of its controversial character, especially when they report how Democritus ridiculed those who enquired into the ἀρχή of the ἄπειρον, frgg. A 39, 65, Diels–Kranz, 5th ed., II, 94, 20 f.; 100, 38 f.

[2] Cf. Ernout-Meillet, *Dict. étymol.*, s.v. Earlier authorities, cf. e.g. W. Prellwitz, *Etymol. Wörterb. d. griech. Sprache*, 2nd ed., 1905, s.v. suggested a relation with Latin *rigere*, to stand straight, but already A. Walde, *Lat. etymol. Wörterb.*, 2nd ed., 1910, 653, regarded this as improbable.

[3] Cl. Baeumker, *Das Problem der Materie*, 1890, 11 f.

It was rather an ordering process by which that which happens 'accidentally', the creation of a cosmos is judged.

It follows from all this that ἀρχή, the beginning or principle, is the point at which the infinite and the finite meet with each other, and consequently is identical for both; and Democritus was quite right to state that such a conception, which changes a vague infinity—an indefinite—into an 'existing' infinite, is much more difficult to apply to the infinite than to the cosmos.[1] The reason for this is that the common contact between the infinite and the finite makes the existence of the infinite an active existence. There may be an 'accidental' creation of a cosmos, but there cannot be an accidental verdict that this cosmos contains ἀδικία, injustice. Neither can it be assumed that man could be the judge in this matter, for such an assumption would at once nullify the 'existence' of the infinite in the process of rational apperception. It is therefore necessary to assume that man is also subjected to the judgment 'in the course of time'. Consequently it appears that for Anaximander the ἀρχή of the world was its judgment by the 'existing', active infinite.

The background against which this theory was evolved is difficult to distinguish, and it is with a certain amount of trepidation that the following attempt at its elucidation is here proposed: I venture to suggest that some of the reports about Orphic and related doctrines, particularly those contained in Damascius', the Neo-Platonist's book *On first principles*, may have a distinct bearing on it. To begin with it may be stated that the word ἀρχή was used as a technical expression in early Orphic poetry. For this we have the witness of Plato to the following verse:[2]

Ζεὺς ἀρχή, Ζεὺς μέσσα, Διὸς δ' ἐκ πάντα τέτυκται.

This verse makes it clear that the word played its part in early Orphic cosmogony. However, the bulk of existing evidence for this cosmogony comes from much later authors, amongst whom Damascius is our most important source. These authors undoubtedly 'modernized' the evidence which they possessed. This is illustrated e.g. by the title of Damascius' work using the plural of ἀρχή. He is, therefore, under suspicion of having inserted the word occasionally when re-modelling earlier Orphic

[1] Cf. *supra*, p. 29, n. 1. The fragments quoted there make it clear that Democritus at any rate credited Anaximander with such considerations as have been proposed here.

[2] Cf. O. Kern, *Orph. frg.*, 1922, 90 f., frg. 21, with references to Plato, *Legg.* IV, 715E, and its scholia.

poetry, especially since it is evident that he did this in his report on the cosmogony ascribed to Acusilaus.[1] There is, however, one fragment preserved by Damascius, where the use of the word is supported by an earlier, and entirely unrelated witness, the Christian apologist Athenagoras.[2] This fragment informs us about the cosmogony ascribed to two otherwise unknown, early Orphic teachers, Hieronymus and Hellanicus.[3] The report once more reminds us of Anaximander, when exhibiting the word *ἀγήρως*, which in later time was remembered as a favourite expression of Anaximander's.[4]

At first sight this Orphic cosmogony does not look very similar to what we have learnt about Anaximander's teaching on 'the beginning'. For it maintains that out of water, and out of that matter which was the chief constituent of earth, a 'third beginning' arose.[5] It seems safe to assume that Anaximander did not propose such a view. It has to be remembered, however, that the description of this 'beginning' as 'the third' is, without doubt, due to Damascius himself.[6] It should, therefore be passed over when an attempt is made at reconstructing the early Orphic doctrine.[7] On the other hand, it may be possible to enlist the help of a very suggestive, if somewhat suspect report by Alexander of Aphrodisias, in which it is maintained that Anaximander's first principle

[1] Cf. Acusilaus frg. B 1, Diels–Kranz, 5th ed., I, 53, 8 f., where the parallel from Philodemus, *De pietate*, and especially Plato's report, *Symp.*, 178B, make it clear that the earlier tradition said nothing here about *ἀρχή*.

[2] Cf. Diels–Kranz, 5th ed., I, 12, 23.—The value of the witness of Damascius lies in the fact that he shows that the rather simple Christian author whose intelligence was hardly sufficient to produce an interpolation which would coincide with an identical one by Damascius, had not been misled, as might well have happened, by an earlier Jewish forgery.

[3] The reasons given by W. K. C. Guthrie, *Orpheus and Greek religion*, 2nd ed., 1952, 85 f., esp. 88, for an early date of this cosmogony are here accepted.

[4] Anaximander frg. B 2, Diels–Kranz, 5th ed., I, 89, 16 f., *ταύτην* (sc. *φύσιν τινὰ ἀπείρου*) *ἀίδιον εἶναι καὶ ἀγήρω*, ct. Damasc. *De princ.* 123 bis, ed. Ruelle I, 317, 15 f. = Diels–Kranz, I, 12, 3, *ὠνομάσθαι δὲ Χρόνον ἀγήραον κτε.*—This parallel has not been mentioned by Guthrie, op. cit.

[5] Obviously Gen. 1:9/10 comes to mind; but it is certain that Damascius, and Athenagoras, were familiar with the LXX so that no far-reaching conclusions about the sources of early Orphism should be drawn.

[6] The reasons for believing this have been set out by W. K. C. Guthrie, op. cit., 79.

[7] It may be remembered, however, that a comparable triad at the origin of the world, Chaos, Gaia, and Eros, is stated by Hesiod, *Theog.* 116 f., where the last might be justly described as 'the third' so long as it is understood that even the *ἔπειτα* in v. 116 does not denote time.

was 'the physics in between air and fire or air and water'.[1] The Orphic theory, in so far as it has been quoted by Damascius, was evidently based upon the conception of a syzygy between the two elements of water and earth; and it would be inviting to derive a similar theory for Anaximander from his undisputed description of matter as a *μεῖγμα*, combined with the, unfortunately rather suspect, report by Alexander.

Much closer relations between Anaximander's theory and this particular Orphic cosmogony can be shown to exist, however, once we turn our attention to that 'third' beginning itself. For this 'beginning' is given the name of Chronos, time, and it is described as a three-headed dragon, whose one head was that of a lion, the other that of a bull, and between them the divine face, and who had wings on his shoulders. It was this Time which was never ageing (*ἀγήραος*), and had for his companions Ananke, necessity, and Adrasteia, recompense. It is this system which I believe to be in close relation to that presupposed in Anaximander's fragment; and if that is accepted, the fragment may be regarded as an apt illustration of the power of abstraction through which the Greek philosopher filtered those oriental symbols,[2] until they had received the clarity of rational, philosophical argument. The dragon Time was translated into Anaximander's 'order of time', the personified Necessity into 'that which is necessary' (*τὸ χρεών*), whilst it was still retained by Empedocles, and the personified Recompense into those earthly ordinances which laid down 'fine and compensation' (*δίκη καὶ τίσις*); and if the 'cosmic egg', the equivalent of Anaximander's conception of 'existence', was not forthcoming from this particular Orphic cosmogony, it was yet produced by the dragon in the closely related 'Rhapsodic cosmogony'.[3]

[1] Anaximander frg. A 16, Diels–Kranz, 5th ed., I, 85, 30 f.—Kranz's terse verdict: 'wrong conclusion drawn from Aristot., *De caelo* III.3, 303b, 12; *Metaph.* I.4, 187, 12 etc.', would not stand if the *μεταξὺ φύσις* could be understood as the relation between the elements rather than as an additional element.

[2] On the Mazdaistic, Persian provenance of the dragon, clearly set out by A. D. Nock, *Harvard Theol. Rev.*, 1934, 53 f., there is no longer any doubt, cf. W. K. C. Guthrie, *Orpheus and Greek religion*, 2nd ed., 1952, 87. The theologian will also notice the close connection with the Merkabah mysticism, and in particular with the visions in Ezek. I, and Rev. 4:7 f.

[3] Damasc., *De princ.* 123, ed. Ruelle I, 316, 18 f. = Diels–Kranz, 5th ed., I, 11, 7 f.—It may also be possible that two other cosmogonies which have been retailed by Damascius, that by Acusilaus, *supra*, p. 31, n. 1, and that ascribed to Epimenides frg. B 5, Diels–Kranz, 5th ed., I, 33, 17 f., were started as analogous attempts at adjusting oriental myths to Greek rational thought. However, so little of these cosmogonies has been preserved that there seems now to be no possibility of making any definite statement about them.

The recognition of these close ties between Anaximander's thought and the Orphic myths about the question of cosmogony is all the more important to us when we try to understand the reason why, right down to the end of the sixth century, no other philosopher, not even Anaximenes who is reputed to have been the pupil of Anaximander, seems to have used *ἀρχή*, meaning 'principle', in his writings; and why it was left to the 'Italian' philosophical school, the school of Pythagoras, to restore the word to its place in the philosophical discussion. For there was no other section of the Greek nation more deeply moved by Orphism than the Greeks of Southern Italy, nor a philosophical school more strongly under its spell than Pythagoreanism. Unfortunately, the sources for the use of *ἀρχή* in early Pythagoreanism are, like the Orphic sources, not easy to handle. There is only one undisputed witness, Philolaus; and apart from him our main evidence comes from Aristotle. He, however, has to be classed as an unsympathetic and at times even hostile witness. In addition to this initial uncertainty, it has to be remembered that the revival of Pythagoreanism in the Neo-Pythagorean school during the Hellenistic period was brought about by men to whom the ideal of scientific honesty had no meaning at all. They were utterly unscrupulous forgers, just as much as, or even more than the Jewish apologists during the same period.

The most important of Aristotle's various reports on the Pythagorean doctrines which is still available is to be found in the first book of his *Metaphysics*.[1] This report will therefore be discussed before going on to the testimonies found in his physical writings. In this report Aristotle turns to the Pythagoreans, after a fairly long reference to the atomistic doctrines of Leucippus and Democritus. The reason for this hysteron-proteron appears to be his intention to deprive the Pythagoreans of the authority which is normally attached to a doctrine because of its earlier date, and which is lost only if and when such a doctrine has been disproved in favour of a later one.[2] Aristotle's purpose becomes clear from

[1] Aristotle, *Metaph.* I.5, 985b, 23 f. = Diels–Kranz, 5th ed., I, 451, 36 f.

[2] There can be no doubt that at the end of the fifth century B.C. the Pythagorean school had passed through a severe crisis which had resulted in a schism between the orthodox and the modernist wings of the school, cf. O. Gigon, *Der Ursprung der griech. Philosophie*, 1945, 123 f. I am attracted by the point of view of B. Farrington, *Greek Science*, Pelican A 142, 1953, 52, that the crisis was caused by the discovery of irrational numbers by some members of the school, first proposed by Tannery, and feel frankly reserved towards Gigon's assertion that the orthodox Pythagoreans were mystics without mathematics.—It has to be stressed, however, that K. Reidemeister, *Das exakte Denken der Griechen*, 1949, 30 f., has shown reasons why Tannery's theory should be regarded with caution.

the way in which he introduces his subject: 'Among these men, and before them, the so-called Pythagoreans, having possessed themselves of mathematics, and being fully trained in it, believed that its *ἀρχαί* were the *ἀρχαί* of the existing universe.' There are two conscious misrepresentations in this sentence. The first, to which attention has just been drawn, is that Aristotle obscured the fact that the Pythagorean school had arisen already in the sixth century B.C. whereas Leucippus as well as Democritus belonged to the fifth.[1] It was therefore untruthful to give the impression that the Pythagorean doctrine arose *ἐν τούτοις*. The second misrepresentation derives from the fact that Aristotle's report starts by comparing the incomparable. The atomists were concerned with causes, *αἰτίαι*, for the functioning of the universe whereas the Pythagoreans enquired into its 'principles', *ἀρχαί*. Aristotle, as the scholar he was, was sufficiently conscientious to change his terminology accordingly; but he wanted his readers to believe that these two different terms had an identical meaning, and in this he succeeded right down to modern time.[2] All the same, it can be proved that his assertion to this effect was wrong, and probably deliberately so. For at the turn from the third to the second century B.C., when the influence of Pythagoreanism—and probably the school itself—had reached its nadir, Theophrastus corrected his master's report. By this time the struggle was over, and thus the disciple calmly reported that it had only been the one Ecphantus 'who represented the Pythagorean monads as corporeal', and thus attempted a compromise between Pythagoreanism and atomism[3]—incidentally obscuring the difference between 'principle', which is imposed upon the cosmos, and 'cause', which is inherent in it.

Aristotle's misrepresentations bring home to us the fact that the connection between Pythagoreanism and the Ionian natural philosophy was only slender. This fact will appear even more clearly when the attitude

[1] Überweg–Prächter, *Philos. d. Altert.*, 12th ed., 1926, 104 f.—W. Spoerri, *Späthellenistische Berichte über Welt, Kultur und Götter*, 1959, 11 f., has lucidly and correctly stated the importance of the distinction between the Greek theory of matter in the sixth and that in the fifth century B.C.

[2] Take for example that Kranz's index to Diels–Kranz, 5th ed., III, 77 s.v. *ἀρχή* makes reference to the article *αἰτία*. F. M. Cornford, *Principium sapientiae*, ed. W. K. C. Guthrie, 1952, 171 f., translates *ἀρχή* in Anaximander as 'the source'. W. Nestle, *Vom Mythos zum Logos*, 2nd ed., 1942, 81, also speaks of 'die letzte Ursache, das Prinzip' (*ἀρχή*), and such examples could be easily multiplied.

[3] Aëtius I.3.15, Diels, *Doxogr.*, 286 = Diels–Kranz, 5th ed., I, 442, 16 f.—The date of Ecphantus is uncertain; but I think that Überweg–Prächter, op. cit., 60, are very likely to be right when making him dependent upon Democritus.

of Democritus to the conception of ἀρχή is examined. At the moment, however, it seems necessary to show that the distinction between 'principle' and 'cause' just made was by no means overlooked by the Peripatetics. That much seems to follow from Theophrastus' report on Parmenides. However, it has to be remembered when this passage is quoted that its text rests upon emendations by modern critics with regard to ἀρχή as well as to αἰτία, which to my mind are both inevitable and convincing, but which may yet carry an element of uncertainty.[1] With this proviso we will now quote Theophrastus:

Of the mixed components [of matter] the central one is the principle and the cause of motion and of coming into being. And this he calls the guiding demon and the holder of the keys, Dike and Ananke.

The two pairs of divine powers mentioned here, the guiding demon together with Adrasteia-Nemesis, the one who 'holds the keys' on the one hand, the Dike and Ananke on the other, are, so it seems, meant to correspond to the pair of 'principle' and 'cause' in the first sentence. If that is accepted it followed that Theophrastus was still aware of the fact that Parmenides had opposed 'principle', which was represented by superior instruction and final judgment, to the mechanics of 'cause', as represented by *equilibrium*, Dike, and by compulsion, Ananke. It is to be assumed that Aristotle too was aware of the existence of this contrast. Therefore he was also aware that the terms of αἰτία and ἀρχή were not unanimously regarded as synonyms by the pre-Socratics; and his report on the atomists and Pythagoreans may be significant for his own tendency of equating these terms,[2] but not for the views of the Pythagoreans. It rather appears that he was not prepared to admit that they were not concerned with cause and causes, but with reasons and principles, and in particular with the supreme *raison d'être* of the universe. To the Pythagoreans the universe appeared as an intellectual, not as a scientific problem. Unfortunately, this fact will now have to be established from a witness who did his best to obscure it.

It had been mathematics, so Aristotle tells us,[3] which had led the Pythagoreans to the conclusion:

that there were many similarities between the simple row of natural numbers and

---

[1] Cf. Parmenides frg. A 37, Diels–Kranz, 5th ed., I, 224, 7 f.

[2] Cf. Aristotle, *Phys.* I.1, 184a, 11, ἀρχαὶ ἢ αἴτια ἢ στοιχεῖα, and *Metaph.* I.2, 982a, 5 etc.

[3] Aristotle, *Metaph.* I.5, 985b, 27 f. = Diels–Kranz, 5th ed., I, 453, 2 f.

that which exists and comes into being, rather than between this and fire, water and earth: So that the one of the numbers be suffering and justice and that [other] one soul and mind, and another one the moment (καιρός), and so also of the others each in its turn. Similarly, observing that which happens in harmonies, and their reasons [expressed] in numbers,[1] seeing that the whole of nature appeared in numbers, and numbers being pre-existent to numbers, they concluded that the elements (στοιχεῖα = chief characteristics) of numbers were elements of the existing universe, and that consequently the entire heaven was harmony and number.

Of course, Aristotle's report does not lay any claim to being an historical account of the origin and evolution of Pythagoreanism. His intention was, in the first instance, to polemicize against the theories proposed by the Pythagoreans of the late fifth and the early fourth centuries.[2] However, since it is the same Aristotle who tells us that Anaximander 'most and above all' regarded number as a constituent of his ἀρχή, the ἄπειρον, 'because it does not have its origin in rational deliberation', i.e. because it is autonomous,[3] and since we have in mind that such had been the Milesian tradition from the time of Thales and his school, who had adopted it from the Phoenicians and the Egyptians,[4] we are faced with the question whether or not Pythagoras derived his conception of ἀρχή as a numerical principle from Anaximander?

There is, I think, some probability for answering this question in the affirmative, even though it might be adduced that Pythagoras had journeyed to Egypt and had thus had the opportunity of drawing from the same source as the Milesian school had done before him. For Pytha-

[1] W. Christ, *Aristotelis metaphysica*, 2nd ed., 1903, 15, has marked this sentence with asterisks, whilst H. Diels has omitted them, but has maintained Christ's punctuation. I believe that the reference to the possibility of expressing harmony by numbers was meant to lead up to the conclusion about the 'chief characteristics' of numbers—rather than to numbers as such—and I have, therefore, in my translation above ventured on a slight deviation from the traditional punctuation.

[2] There can be no doubt that Aristotle did not aim at any of Pythagoras' own writings; and it is even very uncertain whether Pythagoras had ever produced any book, although O. Gigon, *Der Ursprung der griech. Philosophie*, 1945, 124 f., believes it to be probable. The statements by S. Samburski, *The physical world of the Greeks*, 1956, 27, asserting the initial secrecy of the doctrines of the Pythagoreans—as opposed to their ritual—have not been arrived at by a sufficiently critical approach, and should be treated with caution. O. Gigon has sounded a very timely warning not even to put much trust in those legends of which it may be made probable that the real Aristoxenus was their ultimate source. Cf. about him H. Oppermann, *Bonner Jahrbücher*, 1925, 284 f.

[3] Aristotle, *Phys.* III.4, 203b, quoted from Diels–Kranz, 5th ed., I, 85, 26 f.

[4] Cf. Thales frg. A 11, Diels–Kranz, 5th ed., I, 76, 11 f.

goras' journey to Egypt is far from being an established fact. It is true that it is already mentioned by Isocrates; but already in his day such a journey was customarily ascribed to many such fully or semi-mythological figures of the past as Pythagoras, and he has himself fittingly made it appear in a mythological context, in the case of Pythagoras.[1] It is therefore not unreasonable to question the accuracy of Isocrates' report. For since the beginning of the fourth century at least 'the ancient wisdom of the Egyptians' had become the greatest propaganda asset for Egypt's booming tourist industry, and her priests and tourist guides were prepared to confirm the visits of any and all of the foreign celebrities of old. In the sixth century, however, this attraction may not yet have been so strong—and Egypt did not lie on the straight shipping routes from the birth-place of Pythagoras, Samos, to Italy. On the other hand, apart from the possibility of a personal contact between Pythagoras and Anaximander, which however is not too well attested,[2] one may wonder about the adoption of the terms *ἀρχή* and *ἄπειρον* by the Pythagoreans. For these terms are as significant for Anaximander as they are un-attested for Pherecydes, the reputed master of Pythagoras.[3] I feel that the terminological analogies outweigh in importance Porphyry's, the Neo-Platonist's, assertion that Pythagoras drew for his geometry on the Egyptians, for his arithmetic on the Phoenicians, and for his astronomy on the Chaldaeans, which may be no more than an enlargement of the exactly identical report, which the same author has given us about Thales.[4] As a statement of fact it seems to be true that

[1] Isocr., Busiris 28, cf. Diels-Kranz, 5th ed., I, 97, 24 f.—I cannot agree with O. Gigon, op. cit., 129, who holds that such a journey was already 'hinted at' by Herodotus II.81. More credulous still is S. Samburski, *The physical world of the Greeks*, 1956, 26, who claims that Pythagoras 'spent many years' in Egypt and in Babylon. No reputable evidence for this can be adduced. Even Hippolytus, *Philos.* I.2.12, Diels, *Doxogr.*, 557, who himself is only a poor witness, does not say any more than *πρὸς Ζαράταν τὸν Χαλδαῖον ἐληλυθέναι τὸν Πυθαγόραν*, cf. on this assertion J. Lévy, *Recherches sur les sources de la légende de Pythagore*, 1926, 44n. 5; 82. On the spuriousness of the sources attesting the journey to Babylon, esp. *Theol. arithm.* 40; Jambl., *Vita* 19, which arose out of the witches' cauldron of the Neo-Pythagorean forgers, cf. id., 75 f.

[2] Cf. *supra*, p. 28, n. 2.

[3] It may well be that *τὸ ἄπειρον* was yet another new introduction by Anaximander into the philosophical terminology. Hesiod, it is true, uses the word *ἀπείριτος*, cf. Liddell-Scott s.v., but according to the same Liddell-Scott, *ἄπειρος* does not appear in non-philosophical speech before Herodotus and Pindar, at the beginning of the fifth century.

[4] Porphyry, *Vita* 6, Diels-Kranz, 5th ed., I, 100 f., cf. *supra*, p. 36, n. 4.

the school of Pythagoras was under a real obligation to the nations mentioned in respect of the subjects enumerated by Porphyry;[1] but it should be obvious that this recognition does not provide us with a clue to the life of its founder.

The reserve with regard to the theory of Pythagoras' own immediate dependence upon Near-Eastern, and in particular upon Egyptian influences is strengthened by the fact that his theological doctrine of the migration of souls was likewise ascribed to Egyptian influences, a thesis which was proposed already by Herodotus, and was repeated for over 900 years, right to the end of the Neo-Platonic school.[2] For it can be stated with certainty that it is impossible to maintain such a claim because the doctrine of the migration of souls was entirely alien to Egyptian religion.[3] This observation is, I believe, significant for the tradition of the entire body of Pythagorean teaching. For the doctrine of the migration of souls had been representative for Pythagoreanism since the time of its founder, at any rate in so far as his ethical teaching was concerned. And since the ethical and the physical teaching of the school were inseparable, as Aristotle testifies,[4] the proof that its most significant ethical tenet was falsely ascribed to Egyptian influence must make us suspicious also of similar ascriptions in the physical field. It is well known that Pythagoras met with much opposition. Heraclitus, for instance,

[1] B. L. van der Waerden, *Die Astronomie der Pythagoräer*, Verh. d.k. nederl. Akad. van Wetens. XX.I, 1951, 27 n.I, has shown Babylonian precedents for Pythagorean astronomical observations, and on p. 10, has also announced a proof for the dependence of Pythagorean arithmetic upon Babylonian arithmetical teaching. He has, ibid., supported Neugebauer, *Studien zur antiken Algebra* III, Quellen und Studien zur Gesch. der Mathematik, B.3, 245, 'that Greek geometrical algebra issued forth from Babylonian algebra by geometrization of its proofs'.

[2] Cf. Diodorus Sic. I.98; Philostr., *Vita Apol.* III.15 (109); Zacharias Scholast., Th. Hopfner, *Fontes relig. Aegypt.*, 1922–, 678 etc., all dependent upon the report on the teaching of the 'anonymous' Pythagoras in Herodot. II.123.—It is because of this anonymity, as opposed to the mention of Pythagoras' name in Herodot. II.81, that I believe in the secrecy of the Pythagorean ritual, which was meant to safeguard the migration of the soul, and possibly its deliverance, as opposed to the Pythagorean physical doctrines. A short description of such a Pythagorean ritual, celebrated as a protection against mortal danger at midnight, may be found, however, in Philostr., *Vita Apol.* III.93.

[3] Cf. W. M. Flinders-Petrie in Hastings, *Encl. Rel. Eth.* XII, 431 f. The same point was made already by L. v. Schroeder, *Pythagoras und die Inder*, 1880, a little book which even today may claim some interest.

[4] Aristotle, *Metaph.* I.5, 968a, 22 f., together with Magna mor. I, 1182a = Diels-Kranz, 5th ed., I, 452, 35 f., and ibid., 20 f.

accused him of plagiarism.[1] It may well be, therefore, that the Egyptian origin of his doctrine was first adduced in order to substantiate such an accusation, particularly if his connections with his Ionian predecessors had not appeared sufficiently damaging to his fame. Herodotus, on the other hand, maintained that in Egypt flowed the spring of wisdom, and on this belief Isocrates' report about the Egyptian visit of Pythagoras was apparently founded with the intention of proving him to have been truly wise. An exclusive dependence of Pythagoras upon non-Greek sources for his philosophy cannot be shown by reference to these legendary tales.[2]

The doctrine of the migration of souls belonged, on the other hand, already from an early time to the main beliefs of Orphic religion; and whether the Orphics had adopted it from Pythagoras or, as I think probable, Pythagoras had adopted it from them, this common conviction shows clearly the close relations between Pythagoreanism and Orphism already at the end of the sixth century.[3] Since we have previously stated our reasons for assuming that an equally close relation existed between the Orphic cosmogonies and Anaximander's teaching on *ἀρχή*, and since Orphism was an exclusive sect, however much its beliefs may have been 'in the air' during the sixth to the fourth centuries B.C., I venture to conclude that the adoption of Anaximander's terminology of *ἀρχή* and *ἄπειρον* already in early Pythagorean teaching, was due to Pythagoras' own dependence upon Anaximander. This hypothesis is meant to specify the assumption frequently made that Pythagoras was generally influenced in his doctrines by the Milesian school of natural philosophy.

If, therefore, on general grounds the dependence of Pythagoras upon Anaximander's doctrine can be assumed, according to which number was the chief among the components of that *ἄπειρον* (which Anaximander regarded as the first principle, *ἀρχή*) in which the indefinite

[1] Heraclit. frg. B 81, Diels–Kranz, 5th ed., I, 165, 9, cf. O. Gigon, *Der Ursprung der griech. Philosophie*, 1945, 224.

[2] How difficult it is to resist such a tendency may be shown by the example of I. Lévy, *Recherches sur les sources de la légende de Pythagore*, 1926, 44 n.6. He reports that Diog. Laert. VIII.14, says that Pythagoras introduced weights and measures, and then proposes to enlarge this report by the words 'from Egypt', on the strength of Hippolytus' (!) testimony, *Philos*. I.2.18.—The tendency of making Pythagoras an Easterner from Tyre (Porphyry *Vita* 1–2) goes back at least to Neanthes, 'without doubt the historian of the beginning of the 3rd century B.C.', I. Lévy, op. cit., 60.

[3] This date is assured by the almost certain assumption that Pindar, Ol. 2, was written under Orphic influence. For further evidence cf. W. Stettner, *Die Seelenwanderung bei den Griechen und Römern*, 1934, 20, and *passim*.

received the existence of a definite infinite as opposed to the finite, an existence of a circle or—more probably—a sphere,[1] the question has then to be asked why it was that already the early Pythagoreans changed from the one first principle, *ἀρχή*, to several first principles, *ἀρχαί*. The scarcity of reliable sources for the early history of Pythagoreanism makes it impossible to say whether or not this had been Pythagoras' own change of Anaximander's doctrine. The fact, however, that the Pythagoreans of the late fifth and early fourth centuries B.C. spoke of *ἀρχαί* is well supported by our sources. 'Proclaiming the numbers to be *ἀρχαί*', says Aëtius, who for his information was under obligation to Theophrastus;[2] and his witness also receives strong support from a passage in Aristotle's *Metaphysics* quoted above.[3] In his *Physics*, however, Aristotle draws a different picture. There he maintains that the Pythagoreans knew of no more than two *ἀρχαί*, the *ἄπειρον* and the *πεπερασμένον*.[4] Thus our first task will be to show that these two reports are not mutually exclusive. Only when this has been done shall we be able to present certain conclusions about the Pythagorean doctrine of *ἀρχή*.

The two statements by Aristotle just quoted obviously differ in one respect: the one refers to the numerical character of the Pythagorean first principles, whereas the second refrains from making any such express statement. It can be shown, however, that this difference is only superficial. The fact is that Aristotle in his *Physics* contented himself with a very abbreviated formula. That much follows clearly from the continuation of his report on Pythagoreanism in his *Metaphysics*. There he states:[5]

The Pythagoreans, however, taught the existence of two principles in the same manner. But they added a great deal (*τοσοῦτον*) which is their own property, namely that the infinite and the one are not by nature different from fire or earth

---

[1] K. Reidemeister, *Das exakte Denken der Griechen*, 1949, 94 f., has quite rightly stated that Anaximander regarded the cosmos as sphaeric, but has described the *ἄπειρον* simply as 'boundless', neglecting the fact that in so far as it is determined by its contrast to the cosmos it must also be sphaeric, if concave, and beyond that man cannot reason. Empedocles' *ἀπείρων σφαῖρος* seems to mean this very thing, cf. *supra* p. 24 n. 3.

[2] Aëtius I.3.8, Diels, *Doxogr.*, 280 = Diels–Kranz, 5th ed., I, 454, 35 f.

[3] Cf. *supra*, p. 35, n.3.

[4] Aristotle, *Phys.* III.4, 203a, 1 f.; III.5, 204a, 29 f., quoted from Diels–Kranz, 5th ed., I, 459, 10 f., 28 f.

[5] Aristotle, *Metaph.* I.5, 987a, 13 f. = Diels–Kranz, 5th ed., I, 453, 27 f.

or something like these [i.e. the other elements], but that the infinite as such, and the one as such, are the very substance of those things of which they are predicated, so that number is the substance of the universe.

This statement will have to occupy us for some time. It begins with a very Aristotelian account of what another witness has described as follows:[1]

Pythagoras called the monad, the one of his principles, god and the good, which is the nature of the one, the mind (*νοῦς*) itself; but the indefinite dyad [he called] the demon and evil, around which there is the mass of matter (*ὑλικὸν πλῆθος*).

This assertion, of an uncertain provenance,[2] shows forth a transfer of that early Pythagorean dualism of the *ἄπειρον* and the *πεπερασμένον* into the very different world of Neo-Pythagorean thought. For this reason it is necessary to treat the two personified powers of good and evil, the god and the demon, with the utmost suspicion.[3] For although a philosophy like early Pythagoreanism, which attempted to treat ethics and physics together as one and the same problem, had to consider the contrast between good and evil as a contrast of opposing principles—and there is evidence from Aristotle that, by the end of the fifth century if not earlier, at least one group of Pythagoreans did in fact draw this conclusion[4]—their personification does not seem to be in accord with

[1] Aëtius I.7.18, Diels, *Doxogr.*, 302.

[2] H. Diels, *Doxogr.*, 181, has shown that Aëtius I.7.11–34, as a whole is to be found in its right place in that ordered system, in which the early collections of *opiniones* were arranged. Ibid., 74 f., he has made a strong plea in favour of the view that Aëtius did no more than abbreviate, and in fact presumably gave a reasonable abbreviation, of the traditional accounts of the doctrines of the earlier philosophers. All this I would accept as generally correct, and therefore Aëtius has been understood here as being normally the true, if somewhat muffled voice of Theophrastus. This rule suffers an exception, however, in the cases of philosophical schools still flourishing at the time of Aëtius, as Stoicism or Neo-Pythagoreanism. If Aëtius wished to produce an up-to-date compendium, as I believe he did, he could not neglect results of 'modern research' in the antecedents of these schools. He therefore added in the case of Stoicism the views of recent Stoics, Posidonius in particular, and in that of Pythagoreanism, Neo-Pythagorean forgeries.

[3] This follows already from a linguistic observation: even a philosopher as closely linked with Pythagoreanism as Empedocles used 'demon' and 'god' as synonyms, without finding in the first any connotation of evil. Generally speaking it is very uncertain whether the Greeks before the end of the fifth century B.C. had any conception at all of 'evil spirits', cf. U. v. Wilamowitz, *Der Glaube der Hellenen* I, 1931, 27 f.

[4] Cf. the enumeration of the ten opposing principles in Pythagoreanism, Aristotle, *Metaph.* I.5, 986a, 22 f. = Diels–Kranz, 5th ed., I, 452, 35 f.

the general trend of Greek thought in the sixth and fifth centuries, as we know it. Equally doubtful appears to me the remark about 'the mass of matter' accumulating around the 'evil' dyad.[1] As in the case of god and demon, we are faced with a later interpretation of a much earlier theory. It may well be that this later remark has something to do with the table of opposing principles just mentioned, of which the third position exhibits *ἓν καὶ πλῆθος*. Nevertheless, there can be no doubt of its Neo-Pythagorean origin. However, underlying this interpretation we may find in Aëtius' remark a testimony to the effect that already the early Pythagorean doctrine was a dualistic, and not a monistic one.

The reason why the Neo-Pythagoreans went to such lengths in interpreting their predecessors' doctrine seems to have been that the Pythagorean dualism was not simply of the somewhat shallow matter-and-spirit type. It may even be asked whether Anaximander and Pythagoras themselves had already proceeded as far as stating a general, abstract idea of 'matter'; and there can be hardly any doubt that they had not advanced that far with regard to 'spirit'. It may be held with a fair amount of probability that assertions which are to be found in Aristotle, such as 'the whole heaven is harmony and number', or 'number is the substance of the universe',[2] so far from being scientific statements, show the reflection of an entirely new orientation in theology, which took place amongst the Greeks since the end of the sixth century B.C. It is often averred that the development of the mysteries preceded the physical discoveries of the Greeks, especially in the case of Pythagoras;[3] but it is probably more plausible to assume that the new mathematical and physical discoveries of his time demanded a new appreciation of the divine. For it is clear that the new results of the scientific and mathematical endeavours of early Greek philosophy did not find a scientific system of cosmology, as modern physics found at the beginning of this century. What they did find was a mythological system, which the

[1] Less objectionable, but still very doubtful, is the remark by Alexander Polyhistor in Diog. Laert. VIII.25, *ἀόριστον δυάδα ὡς ἂν ὕλην τῇ μονάδι αἰτίῳ ὄντι ὑποστῆναι*, which should at best only be regarded as a somewhat daring interpretation, because it introduces the conception of 'cause'.

[2] Aristotle, *Metaph.* I.5, quoted after Diels–Kranz, 5th ed., I, 453, 9 and 31.

[3] This theory forms the basis of I. Lévy, *Recherches sur les sources de la légende de Pythagore*, 1926, with its most useful collection of sources concerning the external history of Pythagoreanism, and even O. Gigon, *Der Ursprung der griech. Philosophie*, 1945, 121 f., seems to accept it, if only with a somewhat lesser caution than might be desirable.

philosophers either rejected or transferred beyond—at least—the sub-lunar sphere, either in part or altogether.

Once this intellectual situation of the Greek natural philosophers at the end of the sixth century B.C. is fully understood, there will be less reason for bewilderment when we see how in the fifth century B.C. Empedocles ventured to teach the existence of four elements, whereas Philolaus, on the basis of the discovery of the dodecahedron, demanded that there should be five;[1] or that a convinced member of the Pythagorean sect, Ecphantus, nevertheless attempted to align its doctrine with atomism, as we have seen above. We have to draw one fairly certain conclusion from these two instances, which is that even by the end of the fifth century B.C. the early Pythagoreans had not apparently arrived at a clear-cut definition of matter in general.

It is only from this recognition that we may understand Aristotle's assertion that it was number which, according to the Pythagoreans, was the substance of all the things in this world; and that it was these things which predicated number. For example, it was four, which was real, and apples, which was accidental, and not the other way round. At this point, I believe, we may touch the real root of Pythagorean teaching. This lay in a turn of thought, which led from a recognition that number need not be regarded as a predicate to the imperative that it must not be so regarded.[2] Once this is clearly understood, it will also appear why it was that the One, which was regarded as that which was really eternal,[3]

[1] Cf. the admirable discussion of this event by Eva Sachs, *Die fünf platonischen Körper*, 1917, 41 f. At the same time it seems advisable to me not to pass by the wholesome scepticism of K. Reidemeister, *Das exakte Denken der Griechen*, 1949, 18 f.

[2] Cf. K. Reidemeister, op. cit., 34. It was this doctrine which was opposed by Aristotle, *Phys.* III.5, 204a, 29 f. = Diels–Kranz, 5th ed., I, 459, 28 f.—On the distinction between pure number and numbering number in (Neo-) Pythagoreanism cf. Moderatus in Stobaeus, *Ecl.*, *De arithm.* 9, ed. Wachsmuth I, 21.

[3] It is here that I have to part company with Reidemeister who (loc. cit.) says: 'If the numbers are the first in all nature then, they assumed, the elements of the numbers were at the same time the elements of all things being. These "elements" of the numbers are the even and the odd, the latter the determinate, the former the indeterminate, out of these two arises the One, out of the One arises number, and numbers form the whole heaven.' I find in Aristotle, *Metaph.* I.5, 987a, 15 f., quoted by Reidemeister, op. cit. n.62, the contrast between the One and the *ἄπειρον*, not that of odd and even, and I am puzzled about the logical sequence of his sentence where 'odd' and 'even' are introduced as 'elements' of numbers, and number is only the indirect result of their co-operation. It seems to me that the question about the reason for the One producing number has remained unanswered.

was also seen as the power by which the indeterminate, but potentially finite, was determined as it arose. For if the One is once understood as the substance of anything individual, it follows that it is the substance of all individual things, i.e. *ἓν τὸ πᾶν*. On the other hand, that which is more than one, i.e. the dyad, produces time and time again by accepting the addition of yet another one an infinity of objects (*ἄπειρον*) which, however, even in its sum total, is incapable of reaching the universe under any circumstances. This universe (*τὸ πᾶν*) can be reached only by the One; and this, incidentally, is also the cause why man, along with all other living beings, has to die. For he and they are of the dyad, incapable of combining the beginning with the end: *ἀρχὴ καὶ τέλος, ἓν καὶ πᾶν*.[1] This teaching was closely akin to that Orphic theory which maintained that the beginning, the middle, and the ending were all contained in the great body of Zeus.[2] Here, therefore, the idea as well as the ideal of one, supreme, universal deity made its first appearance in Greek thought; and thus the Pythagorean dualism was defined as a dualism between the deity, which was the determined monad, on the one hand, and the indetermined but, by the intercession of the determined One, determinable dyad, on the other.

When, therefore, Pythagoras and his disciples concentrated upon number as the principle of the 'cosmos' (a word which Pythagoras was reputed to have employed as the first for the description of the universe)[3] it was a theological decision which they took. At the same time his followers were motivated in doing so by their discovery of the fact that numbers follow their own, intrinsic, logical law, not imposed upon them by human fancy, but pre-established independently, a discovery which even Aristotle ascribed to them.[4] They had also, according to the same testimony, discovered that numbers, with their inherent logical law, were reflected by natural phenomena so that these could be adequately predicted before they took place, and could be adequately

[1] Cf. Alcmaeon frg. B 2, Diels–Kranz, 5th ed., I, 215, 4 f. How this was understood by later generations is shown by the great London magical papyrus, K. Preisendanz, *Papyri Gr.Mag.* II, 1931, 26 f. no. VII, col. XVII, exhibiting the Uroboros with the prayer, 'protect my, the N.N.'s body and soul whole and intact'.

[2] Cf. *supra*, p. 30, n.2. The identity of the monad and of Zeus was expressly stated by Xenocrates, the head of the Academy, and successor of Speusippus, 339–315 B.C. In this he was undoubtedly dependent upon Pythagorean doctrines, cf. Aëtius I.7.30, Diels, *Doxogr.*, 304.

[3] Cf. Aëtius II.1.1, Diels, *Doxogr.*, 327, 8 = Diels–Kranz, 5th ed., I, 105, 24 f. Cf. also the remarks by Ch. H. Kahn, *Anaximander*, 1960, 193 f.

[4] Aristotle, *Metaph.* I.5, 985b, 26 f. = Diels–Kranz, 5th ed., I, 452, 1 f.

defined by mathematical formulae. In the simple arithmetical progression of whole positive numbers certain rules could be shown to exist which accounted for more complicated physical phenomena than the accumulation of things co-ordinated with their numbering. The two significant discoveries ascribed to Pythagoras, the geometrical theorem named after him, and the first principles of musical harmony, meant not only the beginning of applied mathematics but, from the presupposition of the reality of number rather than of the natural phenomena, the beginning of experimental science in general.

There can be no doubt that geometers and builders long before the time of Pythagoras had, in order to obtain a right angle, used the simple method of dividing a rope by knots into twelve parts of equal length. When this was formed into a triangle with the side-lengths of three, four, and five, a right angle opposite the five was obtained. I believe that the proof that this was not due to chance, but to a law inherent in these three consecutive numbers, i.e. that the length of the rope was immaterial for the success of the experiment, provided only that the twelve units shown by its knots were really identical in length, has to be regarded as the logical starting point for the assertion that the numbers were the principles and the things only their predicates, and that it was the monad which was divine.[1]

The second great discovery made by Pythagoras was once more connected with the square. String instruments had been in use in Greece and in other countries long before the sixth century B.C. Yet we hear that it was Pythagoras who found out that the height of a note was inversely proportional to the length of the string upon which it was produced, but also that the 'harmonic mean' between the octave, the fourth, and the fifth has its mathematical analogue in the cube with its six faces, eight corners, and twelve sides.[2] These two discoveries from

[1] The report by Diog. Laert. VIII.12, about the discovery of his theorem by Pythagoras himself is, of course, neither much worse nor any better than most of the other reports about him. It is accepted as evidence here because it makes sense historically, whereas its rejection would not, I think, clear up anything. Recently it has been maintained that not only its practical use, but also the theorem as such, which we ascribe to Pythagoras, had been known before him to both Egyptians and Indians, cf. W. Nestle, *Vom Mythos zum Logos*, 2nd ed., 1942, 107 n.8. This interesting statement is of no importance for us as long as it is admitted that the philosophical impact of the theorem upon Greek thought originated with Pythagoras and his school.

[2] Philolaus frg. A 24, Diels–Kranz, 5th ed., I, 404, 26 f. Being both unmusical and totally ignorant of musical theory, I have relied for this inadequate statement upon

which the philosophy of number and harmony, the very essence of Pythagoreanism, seems to have taken its origin, were applied to the world of human experience. An attempt was thus made at explaining, with number as the principle, the system of the world as being simultaneously physical and ethical. The ethical component appears to be the same as in Anaximander's theory of individualization by guilt or, as Empedocles later termed it, by badness. For this reason a ruling principle, *ἀρχή*, in both its meanings of beginning and rule, was demanded, controlling the guilty individual, and offering to him and to the world in which he lived, an harmonious and just existence,[1] thus combining moral and natural philosophy.[2]

These considerations make it clear that, even if Pythagoreanism had as yet no definite conception of matter as such, it had in any case a conception of nature.[3] The Pythagoreans did not, as R. G. Collingwood put it, 'give up the attempt to explain the behaviour of things by reference to the matter or substance by which they were made', but they did indeed 'regard their structure as something of which a mathematical account could be given'. The basic axiom presupposed by Pythagoreanism in its attempt to outline the structure of nature, which is the world in which human life has to be lived,[4] is to be found in Philolaus' significant remark about the place of harmony in the system of nature.[5] Here he maintained that, since the two principles, *ἀρχαί*, the

---

the lucid explanation given by S. Samburski, *The physical world of the Greeks*, 1956, 37 f.

[1] Cf. the polemics in Aristotle, *M. moral.* I, 1182a, 11, quoted from Diels–Kranz, 5th ed., I, 452, 22 f., *οὐ γάρ ἐστι δικαιοσύνη ἀριθμὸς ἰσάκις ἴσος*.

[2] The dialectics of such an attempt with regard to the empirical world are made clear by the catalogue of its ten *ἀρχαί*, as enumerated in Aristotle, *Metaph.* I.5, 986a, 22 f. = Diels–Kranz, 5th ed., I, 452, 25 f., to which reference has been made before. On the number ten, and its significance for the empirical world in the Pythagorean system cf. S. Samburski op. cit., 37, 66.

[3] Here I have to join issue with R. G. Collingwood, *The idea of nature*, 1945, 52 f. Not only do I hesitate to admit his contention that Pythagoras was a pupil of Anaximenes, but I cannot find that the contrast between the void and matter, which is the basis for the definition of matter in the abstract, played any significant part in Pythagorean doctrine. Aristotle's (*Metaph.* I.5, 986a, 15 f. = Diels–Kranz, 5th ed., I, 452, 31) saying *ἀριθμὸν νομίζοντες ἀρχὴν εἶναι καὶ ὡς ὕλην τοῖς οὖσι κτε.*, indicates that it did not. Still less acceptable—and indeed irresponsible—is the statement by H. & M. Simon, *Die alte Stoa und ihr Naturbegriff*, 1956, 40, that 'in the school of Pythagoras *ἀρχή* and *φύσις* were identified'.

[4] In this I largely agree with H. & M. Simon, op. cit., 39 f.

[5] Philolaus frg. B 6, Diels–Kranz, 5th ed., I, 408, 12 f.

monad and the dyad, were of themselves dissimilar, it was harmony by which all things determinate as well as indeterminate, were joined together in the cosmos. For nature—and here I feel a tilt at Ionian natural philosophy with its attempts to reduce nature to one basic element as water, fire, or air—would never become a cosmos if it consisted of identical or even kindred objects only. It has to be admitted that Philolaus was one of the last great genuine Pythagoreans, the teacher of that great Theban, Epaminondas, and a contemporary of Plato; nevertheless, it is clear that Theophrastus treated this view as orthodox Pythagoreanism. For it appears as such in Aëtius' fairly circumstantial report on Pythagoras.[1]

This being so, the second half of Aëtius' report, in which he accounts for the way in which numbers may be thought of as bringing the cosmos into being, also deserves our careful attention. On the one hand, it should be obvious that the question how it could be made plausible that numbers, seemingly the most sterile objects of contemplation, were destined for a creative activity, must have agitated the mind of the Pythagoreans.[2] Philolaus made it clear that they accepted the reality of the cosmos of nature, and did not treat it as a delusion. On the other hand, the sentence now to be discussed[3] can, as it stands, hardly be ascribed to Theophrastus for stylistic reasons. Its terminology at any rate is Stoic rather than Peripatetic, and this becomes almost painfully clear when the sentence following it is read, which in its style agrees well with the familiar diction of Theophrastus. It seems impossible to reject this alien sentence if, as I believe, it is an honest attempt at condensing Theophrastus' report, and not an interpolation of Neo-Pythagorean invention; but even so its manner of presentation raises concern. This is what Aëtius has to say:

According to him [Pythagoras] the one of his principles tends towards the active and form-giving cause, which is Nous, the deity, the other towards the passive and material [cause], which is the visible cosmos.

The question to be asked is, what is the place of cause and effect in a

[1] Aëtius I.3.8, Diels, *Doxogr.*, 280 = Diels–Kranz, 5th ed., I, 454, 35 f.

[2] It seems to be that the acceptance of the sexual connotation of the symbol of the cube in Neo-Pythagoreanism, as described by R. Eisler, *Philologus* 68, 1909, 118 f., 461 f. (cf. my remark in *Harvard Theol. Rev.* 38, 1945, 177 n. 4) has to be understood in the light of this paradox.

[3] Diels–Kranz, 5th ed., I, 454, 39 f., *σπεύδει δὲ αὐτῷ τῶν ἀρχῶν ἡ μὲν ἐπὶ τὸ ποιητικὸν αἴτιον καὶ εἰδικόν, ὅπερ ἐστὶ νοῦς ὁ θεός, ἡ δὲ ἐπὶ τὸ παθητικόν τε καὶ ὑλικόν, ὅπερ ἐστὶν ὁ ὁρατὸς κόσμος.*

system as mathematically constructed as Pythagoreanism, and is there any meaning attached to the rather uncommon use of 'tends towards' (*σπεύδει ἐπί*) in this report?

A little, and somewhat uncertain light is thrown upon this matter by various remarks made by Aristotle which, whilst establishing a link between Aëtius and earlier Pythagoreanism of the fourth century B.C., reminds us of the filter of Peripatetic philosophy through which our passage has passed. Aristotle tells us above all[1] that the *ἀρχαί* themselves are not to be regarded as causes of anything, yet some of the Pythagoreans maintained that their inherent *φύσις* as well as their demanded dialectical opposites, were to be regarded as the causes of nature. This remark not only elucidates the use of 'tends towards' by Aëtius, but also offers a hint as to the working of the aetiological argument in the Pythagorean system. For there can be no doubt that it refers back to the previously quoted table of ten opposites, which Aristotle also ascribed to 'others among them'.[2] We thus learn that Aëtius reported a view which was held by a minority amongst the Pythagoreans, but which was regarded as a permissible interpretation of Pythagoras' teaching—at any rate by Aristotle.

Aristotle's next remark on the subject avers that the *ἀρχαί* according to the Pythagorean doctrine were the causes of plant and animal life, because the highest goodness and beauty are to be found in the *ἀρχή*, and earthly beauty and perfection are to be seen as their products.[3] This doctrine seems to have been closely linked with the Pythagorean teaching about the migration of souls. We know that Empedocles at any rate included plants in the circle to be completed by the soul, and may presuppose the same conviction underlying this report by Aristotle also. However that may be, both Aristotle's remarks serve to indicate what lively discussions were going on amongst the Pythagoreans about the mutual relations between principle and cause. Yet these remarks are insufficient to allow us to state what final results may have been obtained. Only one result is reasonably plain, that the Pythagoreans were intent on keeping the two conceptions of principle and cause apart, a method which Aristotle did not share. Aëtius, on the other hand, can be shown to have preserved a good deal of Peripatetic doctrine, with some admixture of Stoicism, when he condensed Theophrastus' report on the Pythagoreans to the sentence which we have just discussed. For it follows

[1] Aristotle, *Metaph.* XIII.6, 1093b, 7 f. = Diels–Kranz, 5th ed., I, 458, 13 f.
[2] Aristotle, *Metaph.* I.5, 986a, 22 f. = Diels–Kranz, 5th ed., I, 452, 35 f.
[3] Aristotle, *Metaph.* XI.7, 1072b, 30 f. = Diels–Kranz, 5th ed., I, 454, 10 f.

clearly from the perusal of Kranz's index to the fragments of the Pre-Socratics that the terms *αἴτιον ποιητικόν* and *παθητικόν* do not belong to their diction at all, but rather to the terminology of the post-classical, Hellenistic period.[1]

From this point we will have to venture upon even less secure ground, relying almost entirely upon the testimony of the doxographers. Aëtius, our best witness (!), continues: 'For the nature of number is the ten. For up to ten count all Greeks and all barbarians; and having arrived there, they step back upon the monad.'[2] From this saying it follows—and in so far we are still on reasonably safe ground—that the Pythagoreans began their attempt at finding an aetiological connection between pure mathematics and the cosmos of nature by observing the natural mathematics of the common people. This led to the conclusion that such natural mathematics were based upon the decimal system.[3] The aetiological approach, therefore, had to be made from the natural row of numbers, one to ten, although the One was not really regarded as a number by the Pythagoreans,[4] and not from the 'infinite dyad' as such. It appears from the worship of the tetrad in Pythagoreanism that their predilection for triangular numbers may also have had something to do with their decision in favour of the decimal system; and it is an attractive hypothesis that the famous 'Platonic number' of the ideal man was of Pythagorean origin. Whatever may have been the cause of the eclipse of the school by the end of the fourth century B.C., two things seem reasonably certain, the first, that the search for a mathematical formula which could make evident why things happened in the

[1] Diels–Kranz, 5th ed., III, 28, s.v.

[2] Diels–Kranz, 5th ed., I, 454, 41 f., *εἶναι δὲ τὴν φύσιν τοῦ ἀριθμοῦ δέκα, μέχρι γὰρ τοῦ δέκα πάντες Ελληνες, πάντες βάρβαροι ἀριθμοῦσιν, ἐφ' ἃ ἐλθοντες πάλιν ἀναποδοῦσιν ἐπὶ τὴν μονάδα.*

[3] The fact that the 'barbarian' mathematics were consulted without sign of any misgivings, places this testimony in the sixth century B.C. rather than in the fifth or the fourth, cf. my *Politische Metaphysik* I, 1959, 155 n.5. Neither can it be placed in the Hellenistic period, when the 'finger numbering', cf. K. Menninger, *Zahlwort und Ziffer*, 2nd ed., 1958, II, 11 f., which became common at that time, was based upon the centesimal system. This makes me doubt once more the assertion made by S. Samburski, *supra*, p. 37 n. 1, that Pythagoras had sojourned for years at Babylon, where the local 'barbarians' used the duodecimal system. In view of the Pythagorean oath, Aëtius I.3.8, Diels, *Doxogr.*, 282, 8 f., by the tetraktys, which is ancient despite its synthetic Doric language, I feel that it would be regrettable to abandon the view that in number mysticism the decimal system belongs to the (Neo)- Pythagoreans, and the duodecimal to the Chaldeans; but such would be the logical consequence of accepting Samburski's hypothesis.

[4] Cf. K. Menninger, op. cit., I, 30 f.

cosmos of nature, was never abandoned by its various inheritors; the second, that they signified by their pride taken in the introduction of the word 'cosmos' by Pythagoras himself, their conviction that nature could never be regarded as unchained and—even potentially—chaotic.

It is for this reason that I venture to take seriously a report by Aëtius in which we are informed about a Pythagorean theory of cosmogony. It has to be admitted that the reasons against doing so appear rather formidable: First, the report has come down to us only in fragments, and secondly, these fragments themselves are in rather a sorry state. Already the first of them has only got a poor attestation. In the pseudo-Plutarchean collection of *opiniones*, of the second century A.D., we find the bare statement that Pythagoras, Plato, and the Stoics all agreed that the cosmos was created and, but for the divine providence and forbearing, could and would perish.[1] Obviously, this assertion is made rather disingenuously in view of the fact that it presupposes that the three philosophies mentioned had all an identical conception of what was meant by 'the cosmos'. Such a view may have been arguable at the time when pseudo-Plutarch's anthology was compiled, at the end of the Hellenistic period; but we may confidently claim that, in spite of what is said in the *Timaeus*, Platonists and Pythagoreans would have strongly disputed it in the fourth century B.C., as would also Zeno and his followers in the third. We thus have to turn to John Stobaeus, of the early sixth century A.D., who has preserved for us the same excerpt from Aëtius, which forms the basis for pseudo-Plutarch, in a different and, as regards Pythagoreanism, more explicit form[2]. In this testimony Pythagoras and Plato are separated, and the Stoics are omitted altogether. It is, however, not quite easy to vindicate this tradition against pseudo-Plutarch. For John Stobaeus was a Neo-Platonist, and this school was greatly interested in Pythagoreanism without ever attempting to distinguish between early and Neo-Pythagoreanism. This fact renders precarious the support which H. Diels has adduced from the Church historian Theodoretus of Cyrus, and which is in any case counter-balanced by the testimony of Cyril of Alexandria in favour of pseudo-Plutarch.[3] No more can be claimed in favour of John Stobaeus' report than that it existed already by the end of the fourth century A.D.; but we have no external evidence to show what its earlier history may have been. Thus the possibility that it

[1] Ps. Plutarch, *De plac. philos.* II.4.

[2] The identity of Ps. Plutarch's fragment with Stob., *Ecl.* I.20.1, is made evident by H. Diels, *Doxogr.*, 330 f., where the two pieces are printed in parallel columns.

[3] The quotations from the two Church Fathers may be found in Diels, ibid.

is a forgery, although John Stobaeus did not invent it, cannot by any means be ruled out.

On the side of internal evidence there are yet two clear reasons for giving preference to Stobaeus' version, the omission of the Stoics, and also the distinction between Pythagoras and Plato. For the Neo-Platonic school, claiming descent from all the three of them, might well have kept them bundled together. It is also rather unusual, to say the least, to find in the genuine fragments of Aëtius such an en-bloc report, as we have it before us in pseudo-Plutarch. Furthermore, I would maintain that Stobaeus' report agrees much more readily with what is known otherwise about the teachings of early Pythagoreanism than does pseudo-Plutarch's. For Stobaeus states: 'Pythagoras holds that the cosmos is created with regard to design (*ἐπίνοια*), though not with regard to time.' This saying, we can see at once, is not a 'Pythagorean' quotation. It may well come from a Peripatetic source—Theophrastus—but it would, perhaps, be more cautious to say that it is not really characteristic of any particular philosophical school.[1] It is remarkable, however, that this saying might well serve as a summary of the Pythagorean doctrine which maintained that the whole heaven consisted of numbers. I feel that some such statement is even logically demanded once the philosophical system of the early Pythagoreans has been clearly grasped, that numbers are the principle (*ἀρχή*), and that the natural phenomena are their concomitants; and it is for this reason that a certain amount of confidence may be given to Stobaeus' report.

The second fragment of information on Pythagorean scientific, cosmological speculation also comes from Aëtius,[2] and is perhaps even more difficult to reconstruct than the first. It is certain that Aëtius and, perhaps, already Theophrastus had arranged their anthologies according to subjects. The second book of Aëtius' *opiniones* dealt with the cosmos, its extent, its shape, spirit, duration, the supply of its constituent materials, and in the sixth chapter with the problem, which of the elements had held the position of *ἀρχή* in the creative activity of the deity when establishing the cosmos. It is evident that such a question could hardly arise in the context of early Pythagorean doctrine, not even if it were true, as Aëtius seems to have asserted, that it had exercised the minds of

[1] I have not found the contrast *κατ' ἐπίνοιαν—κατὰ χρόνον* in the Pre-Socratics, in Plato, in Aristotle, or even in v. Arnim's fragments of early Stoicism. The Stoic complement, rather than contrast, is *ἐπίνοια—περίπτωσις*, cf. H. v. Arnim, *S.V.F.* II, 29, 25 f.

[2] Aëtius II.6.1, Diels, *Doxogr.*, 333.

contemporary 'natural philosophers'. For he assures us that these had held the view that the first element to be used had been earth, because 'it is the centre that is the ἀρχή of the sphere'.[1] This sentence is followed by the brief remark that Pythagoras had held that 'fire and the fifth element' had constituted the first beginnings of creation.[2] This information appears to be even more suspect than the preceding remark. Its attestation by both pseudo-Plutarch and John Stobaeus is impaired by the fact that Galenus, who in this place seems to follow Aëtius faithfully in all other respects, nevertheless omitted the fifth element.[3] It is also highly improbable that Pythagoras, who seems to have known only four—and perhaps only three—regular bodies, the cube, the pyramid, the octahedron for certain, and probably also the icosahedron,[4] and may well have aligned those four with the commonly accepted four elements, should have thought of a fifth element at a time when the fifth regular body, the dodecahedron, had not yet been constructed, let alone been proved to be the final one.

These considerations have been prompted by the fact that Aëtius, after inserting accounts of the views of Empedocles and Plato on the earliest of the elements, seems to have returned to Pythagoras once more when, in the second half of our fragment, he expounded the doc-

[1] Aëtius II.6.1. It is Ps. Plutarch, *De plac. philos.* II.6, who is responsible for ascribing this view to the φυσικοί, whereas Stobaeus ascribes it to the Stoics; and it is H. Diels, *Doxogr.*, 333, who treats Ps. Plutarch as authentic. I have not dared to question Diels' great authority, though with some hesitation. The reason for this is the fact that Ps. Plutarch himself (op. cit., II.2 = Aëtius II.2.1 f., Diels, *Doxogr.*, 329) merely mentions the Stoics as maintaining that the cosmos is of a spheric shape, whilst the 'others' whom he mentions as describing it as a cone, or as egg-shaped, are the Ionian Anaximander, and the Orphics. On the other hand, the evidence for the atomists, Leucippus and Democritus, believing in a spheric cosmos, is only supplied by Stobaeus, who nevertheless ascribes the principle quoted to the Stoics alone, who are also the most likely school to have formulated it. Diels's very strong reason for deciding in favour of Ps. Plutarch is, however, that Aëtius, wherever we come to grips with him, arranged his material historically under its various headings. He would, therefore, not have started with the Stoics, as he seems to do in Stobaeus, if earlier material was at hand.

[2] So Aëtius II.6.2, Diels, *Doxogr.*, 333, 19 f.

[3] Galen., *De plac. philos.* 49, Diels, *Doxogr.*, 622, 9.

[4] Cf. *supra*, p. 43, n. 1. The thesis that Theaetetus discovered both the icosahedron and the dedecahedron, cf. S. Samburski, *The physical world of the Greeks*, 31 f., rests solely upon Suidas s.v. Theaetetus; and whilst K. Reidemeister, *Das exakte Denken der Griechen*, 1949, 18 n.8, may overemphasize the unreliability of the sources of the Byzantine lexicographer, he has convinced me at least, 'that Theaetetus the mathematician is no more than a legend', ibid., 26.

trin of the concomitance of the regular bodies with the elements.[1] It is thus strongly impressed upon us that those Pythagoreans of whom Theophrastus had known, or at least some of them, ascribed to the entire cosmos the shape of a dodecahedron, whilst claiming that its creation had begun from the simplest of the regular bodies, the pyramid, and its concomitant element, fire. This, it seems to me, has a genuine Pythagorean ring, especially since it is well known that the Pythagoreans did not represent earth as the centre of the universe, but constructed the cosmos around a central fire.[2] Nevertheless it would be hazardous to prefer for this one reason the report of Galenus mentioned above, which combines the remarks about the 'first' element with those about the concomitance of the elements with the regular bodies, to those of pseudo-Plutarch and John Stobaeus upon which H. Diels has based his reconstruction of Aëtius. All that can be held is that Galenus has preserved in his report some genuine Pythagorean doctrine.

I feel more doubtful still with regard to the thesis of S. Samburski, who assumes that the Pythagoreans regarded the aether as the fifth element. It is true that Aëtius' intervening remarks about Empedocles, as reported by pseudo-Plutarch, when combined with Philolaus' description of the dodecahedron as the 'vessel' (ὁλκάς) of the cosmos, may lead to such a conclusion; but it is historically impossible. The theory became very popular with the Hellenistic philosophers; but no earlier witness for it can be found than the pseudo-Platonic *Epinomis*. Even more significant is the fact that Plato's *Timaeus* does not show the slightest trace of knowing it. For it may confidently be held that this book was under greater obligation to Pythagoreanism than, perhaps, any other of his writings. Finally, Aristotle makes it quite clear that the theory is only of recent origin when he says: 'Therefore, since the first body is something different from earth, and fire, and air, and water, they (the earlier writers) have given the name of aether to the supreme place (the place of the gods) etc.'[3] All this makes it appear very doubtful

[1] Aëtius II.6.5, Diels, *Doxogr.*, 334 f.

[2] Cf. Chr. A. Brandis, *Gesch. der Entwickelungen der griech. Philosophie*, 1862, 181 f. B. L. van der Waerden, *Die Astronomie der Pythagoräer*, Verhand. d. K. Nederl. Akad. van Wetens., afd. I Natuurk. XXI.1, 1951, 49, ascribes to Pythagoreanism the geo-centric system as it is found in Plato's *Timaeus*. However that may be, I believe that the system based upon a central fire, which v. d. W. discusses on p. 49 f., was the original Pythagorean conception. Cf. Aristotle, *Metaph.* XIII.6, 1080b, 20; XIV.3, 1091a, 13 f.

[3] Empedocles, cf. Aëtius II.6.3, Diels, *Doxogr.*, 334, 1 f.; Philolaus frg. 12, Diels–Kranz, 5th ed., I, 413, 2; Ps. Plato, *Epinomis*, 981C, 984B; Aristotle, *De caelo* I.3, 270b, 20 f.

that Empedocles, as Aëtius reports, really held the view that the aether was the first element which was separated from the infinite sphere;[1] and it is equally doubtful whether the Pythagoreans held such a view—even if it is believed that they taught about a 'highest aether' as the supreme and ultimate sphere, the sphere of the immortals.[2]

The reason why I am rather tempted to find in Galenus' report a pointer to the real Aëtius, and through him to the view held by Theophrastus, is that it seems to find support in a well-attested saying of Philolaus'. Galenus said:

> Pythagoras presaged the regular bodies of the third dimension, which he calls the 'mathematical' bodies. For out of the cube, he says, the earth has gone forth, out of the pyramid the fire, out of the octahedron [* * *] the water, out of the dodecahedron the sphere of the universe.

Philolaus, on the other hand, proclaims:

> And thus the bodies of the sphere are five, those inside the sphere as fire and water and earth and air, and the vessel (?) of the sphere the fifth.[3]

A comparison of the two sayings will make it clear that the source of Galenus—Theophrastus–Aëtius, as I believe—had done no more than to give an elucidation, which necessitated a slight enlargement, of the somewhat obscure saying of Philolaus. As an introduction, however, Galenus and his source stated that Pythagoras 'presaged' (ᾠήθη) the five regular bodies, and in doing so they have advised us that he was believed to have postulated their existence even before his pupils succeeded in constructing them, and in establishing their mathematical principles. If such a conclusion is valid it will serve to show how the Pythagorean cosmogony, right from its earliest beginnings, was built upon 'number' in its widest sense commanding the concomitance of natural phenomena in the following way. Because the abstract figure of the pyramid (which in this way has the quality of 'number') finds its representation in the world of sense-appearances in the element of fire, and likewise the cube in the element of earth,[4] therefore two more regular bodies, the repre-

[1] It would be possible to reject Aëtius' thesis altogether if Kranz's translation of Empedocles frg. B 38, Diels–Kranz, 5th ed., I, 328 f., were certain; but such is not the case. Cf. also Empedocles frg. B 53, Diels–Kranz, 5th ed., I, 332, 11 f.

[2] Diog. Laert. VIII.26 fin. = Diels–Kranz, 5th ed., I, 449, 15 f.

[3] Galenus, cf. *supra*, p. 52, n. 3. Philolaus, cf. *supra*, p. 53, n. 3, and the discussion of his views by S. Samburski, op. cit., 32 f.

[4] Cf. my article 'Vir bonus quadrato lapidi comparatur,' *Harvard Theol. Rev.*, 1945, 177 f.

sentation of which is established by the experience of the elements of air and water, are demanded and subsequently discovered. The discovery of a fifth 'mathematical' body, however, constituted a considerable inconvenience for the Pythagorean system, since no further evident experience of the appearing world could be adduced as its representation.

4 The Pre-Socratics of the fifth century B.C.

PARMENIDES

It has thus become evident that for the Pythagoreans the term of ἀρχή was a fundamental one for their idealistic philosophy. By it they were enabled to maintain simultaneously the eternity of an intellectual world of numbers and geometrical figures as well as the creation of the material world of human experience—possibly together with its final destruction, although we hear nothing of this in our, admittedly rather insufficient, sources. This approach enabled them also to develop a system of ethics, based upon the migration of immortal and imperishable souls. From the many features which they had in common with Orphism and the philosophies of Anaximander and Empedocles, it follows that for them the origin of the material world was bound up with a 'fall' of the soul; but there seems to be no evidence for the assumption that this fall was seen by them in any way as a pollution of the soul by matter. For it seems at any rate clear that for Empedocles the polluting material was rather the blood of sacrificial victims, but neither killing nor matter as such; and as yet we know far too little about the religious, ethical, and philosophical convictions of early Pythagoreanism to be able to reject this analogy. It will be a task for future scholars to decide whether or not an abstract idea of matter was formed by the early Pythagoreans, and whether this was done in order to give a logically satisfactory explanation of what they meant by the pollution of the soul, employing for this purpose the contrast of matter and spirit.

It is also clear that any doctrine of the fall of the soul will be turned into an esoteric doctrine, as it leads to the demand of salvation; and it was this fact which caused the other Pre-Socratics to turn against Pythagoreanism. This was made most clear in the attitude of Heraclitus.[1] However, it seems to me advisable to relegate Heraclitus to the chapter dealing with Stoicism, whose spiritual ancestor he was claimed to be,[2]

[1] Heraclit. frgg. B 40, 81 (the historian Timaeus' report of the conflict 121), Diels–Kranz, 5th ed., I, 160, 169, 180 f. The fragment B 129 = Diog. Laert. VIII.6, which might also be quoted in this context, is spurious, according to Kranz's verdict. I am inclined to agree.

[2] M. Pohlenz, *Die Stoa* I, 1948, 68.

although in one of the fragments of his work the term *ἀρχή* is used in a mathematical sense.[1] For a different reason it seems impossible to discuss with profit the views of Protagoras, although he was credited by the doxographers with a book called *Περὶ τῆς ἐν ἀρχῇ καταστάσεως*. For of this book only the title, but not a single word from its text seems to have survived. It is therefore impossible to decide what sort of a constitution (*κατάστασις*) the great Sophist may have had in mind, that of the cosmos or that of the city-state. For although it is true that the Greeks held that there existed a close analogy between these two,[2] it would yet make a distinct difference whether the creation of the city-state or the creation of the world was the subject of his examination.

It is evident, on the other hand, that fifth-century Greek philosophy took up the term *ἀρχή* with a new intensity, but also from a new angle. It is at least probable that the causative meaning of the term made its entry into pre-Socratic philosophy at this time; and it is to be assumed that the man who introduced it was Parmenides. Admittedly, in the extant fragments of his work the word as such does not appear. What does appear, however, is an evidently polemical remark attacking, so it seems, the Pythagoreans. The relation between Parmenides and the Pythagoreans was indeed ambiguous. For, like them, Parmenides also distinguished between a world of appearances and the world of reality, the world of IS. Of this world he proclaimed the following views:[3]

> *αὐταρ ἀκίνητον μεγάλων ἐν πείρασι δεσμῶν*
> *ἔστιν ἄναρχον, ἄπαυστον · ἐπεί γένεσις καὶ ὄλεθρος*
> *τῆλε μαλ᾽ ἐπλάχθησαν, ἀπῶσε δὲ πίστις ἀληθής.*

(Again it is immovable within the bonds of great fetters, without beginning, without ceasing; for creation and perdition are beaten far back: True conviction has banished them.)

[1] Heraclit. frg. B 103, Diels-Kranz, 5th ed., I, 174, 1 f., *ξυνὸν γὰρ ἀρχὴ καὶ πέρας ἐπὶ κύκλου περιφερείας*. I agree with W. Kranz, against Wilamowitz, *Hermes*, 1927, 276. The primary interpretation of the sentence is purely geometrical, and it is at least doubtful whether a metaphorical application to the philosophical problem of infinity was at all intended.

[2] Cf. my *Politische Metaphysik* I, 1959, chap. IV, 1.

[3] Parmenides frg. B 8, 26 f., Diels-Kranz, 5th ed., I, 237, 6 f. The fragment is interpreted here as an anti-Pythagorean proclamation in accord with J. Raven, *Pythagoreans and Eleatics*, 1948, 21 f. Our concentration upon the term of *ἀρχή* enables us to claim that a special tilt in this direction is to be found in the word *ἄναρχον*, a point which Raven seems to have missed.

For the interpretation of these lines, J. Raven has rightly referred to a critical remark by Aristotle, based upon Parmenides, in which he censured the Pythagoreans, because their ordering principle, number, left untouched the question of motion and coming into being:[1] 'They do not say a word how it should be possible that without motion and change the coming into being and perdition should take place.' Now the whole essence of Parmenides' repeated protests is just this that the cosmos of IS, of true reality, which is to be reached on the road of truth, has never come into being, and is incapable of change and perdition (*ὡς ἀγένητον καὶ ἀνώλεθρόν ἐστιν*).[2] This, it has to be stressed, is for him also an eschatological protest. Dike was not seated by Parmenides at any odd gate; but when he approached her, at the beginning of his poem, he found her sitting at the gate of day and night, at the gate of Hades.[3]

Parmenides, therefore, like Anaximander and Empedocles, was moved by the recognition that his position in this finite world of appearances, his individualization, was a matter of eternal justice and injustice. It is not always clearly recognized that after Parmenides had been shown by Dike the choice of the two roads, he did not take either of them, but descended again to his fellow-men. He did not claim—and in this he was clearly distinguished from Empedocles[4]—to have been deified, and to have entered the world of true reality, the cosmos of IS, already. In fact, it would appear that the second half of his poem consisted of a philosophical appreciation of his descent back into earthly life. It was for this reason also that Parmenides approached this world of appearances in a spirit which is so very different from that of the Pythagoreans. He evidently rejected their theory of the concomitance between number and its *φύσις*, a theory which they had indeed adopted on somewhat too easy terms; and his reason for doing so was not only that no rational law can be established which would result in such a co-ordination, but

[1] Aristotle, *Metaph.* I.8, 990a, 10 f. = Diels–Kranz, 5th ed., I, 456, 26 f., cf. J. Raven, op. cit., 35 f.

[2] Parmenides frg. B 8, 3, Diels–Kranz, 5th ed., I, 235, 4.

[3] Parmenides frg. B 1, 10–14, Diels–Kranz, 5th ed., I, 229, 6 f., cf. my 'Dike am Tor des Hades', *Scritti in mem. E. Albertario* II, 1950, 549 f. The recognition of the eschatological character of Parmenides' poem is essential also for the correct interpretation of Heraclit. frg. B 15, Diels–Kranz, 5th ed., I, 154, 18 f., a fragment which K. Riezler, *Parmenides*, 1934, 29, has rightly adduced for the illustration of Parmenides' intellectual situation.

[4] Empedocles frg. B 112, 4, Diels–Kranz, 5th ed., I, 354, 17, *ἐγὼ δ' ὑμῖν θεὸς ἄμβροτος οὐκέτι θνητός*.

was in particular the clear recognition that any such theory made light of the pre-eminence of the One. No natural process but reasoning (νοεῖν) was required in order to establish the relation between the true cosmos of IS, and this world of appearance (δόξα). For if such were not the case, if the Pythagorean eschatology with its continuous and, as it seems, painless migration of souls were accepted, the judgment inherent in the prophet's rather than philosopher's instruction by Dike[1] would be meaningless.

With Parmenides, therefore, we reach at last the question of what is exactly the product of creation? His answer is that it is the non-existent, a fallacy. That which really exists cannot have come into being by an act of creation because it is truly spiritual. This is the meaning of that most challenging and most puzzling of all Parmenides' sayings: 'For the same is to think and to be.'[2] When this is read the first reaction is a comparison with the Cartesian statement *cogito ergo sum*. The question has to be asked whether the philosopher from Elea like Descartes only meant to establish by his saying the human existence? It seems to me that the context of the saying offers no help for its interpretation because its place in the Parmenidean poem is quite uncertain.[3] Under these circumstances the way in which the saying was quoted by its two witnesses, Clement of Alexandria and Plotinus, assumes a special significance. Clement joined it to a fragment from an unknown comedy by Aristophanes, saying: 'For the same power as the seeing has the thinking.'[4]

[1] It seems to be a tradition in recent research not to identify 'the goddess' in Parmenides frg. B 1, 22, with Dike who is mentioned eight lines earlier. W. Nestle, *Vom Mythos zum Logos*, 2nd ed., 1942, 112, calls her 'truth', and O. Gigon, *Der Ursprung d. griech. Philosophie*, 1945, 246 f., even characterizes her as a relation of the Muses, as in Hesiod and Empedocles. The identification with Dike rests upon the testimony of Theophrastus, cf. Parmenides frg. A 37, Diels-Kranz, 5th ed., I, 224, 3 f. O. Gigon's suggestion in particular seems to make light of the fact that Parmenides was opposed to both Hesiod and Empedocles.

[2] Parmenides frg. B 3, Diels-Kranz, 5th ed., I, 231, 22, *τὸ γὰρ αὐτὸ νοεῖν ἐστίν τε καὶ εἶναι.*

[3] W. Kranz's dictatorial: 'To be joined to B 2', has only the value of an hypothesis which is by no means self-evident. For unless it is maintained that frg. B 2, 7 *γνοίης*, should be understood as being synonymous with *νοεῖν*, an assumption which seems rather far-fetched to me, no obvious link between the two fragments can be shown. Still more misleading are earlier theories like that of C. A. Brandis, *Gesch. d. Entwickelungen d. griech. Philosophie*, 1861, 88 n.29, based upon Mullach's fanciful reconstructions.

[4] Clem. Alex., *Strom.*VI.23, quoted by Diels-Kranz, loc. cit., *δύναται γὰρ ἴσον τῷ ὁρᾶν τὸ νοεῖν.*

The fact that this parallel was culled from a current anthology by Clement can hardly be doubted; and this means that it represents an interpretation which was traditional in Alexandrian popular philosophy. From this it follows that Plotinus, who was also an Alexandrian, and presumably familiar with the popular interpretation of Parmenides' saying, expressed his own personal thought when he stated: Parmenides 'combined as identical existence and the Nous'.[1] Considering the philosophical qualities of our two witnesses, it seems tempting to give preference to Plotinus' interpretation of our fragment. Against this preference there arises, however, a serious obstacle from the use of the word Nous in the extant fragments of Parmenides' poem. For it is true to say that the Nous is understood as the ability to think either in the right way, as Parmenides is exhorted to do,[2] or in the uncertain fashion of mortal men.[3] 'For just as each man possesses the mixture of erring members (the senses), so also does the Nous accompany him.'[4] Thus the Nous is seen as a corrective of human error, and it therefore appears that Clement's analogy from Aristophanes has much to commend it. For it is evident that the parallel between the sense-perception and the Nous, the first misleading, and the second correcting, was drawn already by Parmenides himself.

Having stated this, it is nevertheless impossible to reject Plotinus' assertion as being no more than a Neo-Platonic invention. What has become open to doubt is no more than that Parmenides regarded the Nous as the basis of all that really exists. For there are certain ancient testimonies which make it seem more likely that he rather regarded the Logos as such a basis.[5] If this is accepted as the correct view it would make it clear how, by way of the common Hellenistic identification of the Nous with the Logos, Plotinus came to his misleading interpreta-

[1] Plotin., *Enn.* v.1.8, quoted from Diels–Kranz, 5th ed., I, 231, 17 f.—It is to be regretted that both K. Riezler, *Parmenides*, 1934, 69 f., and O. Gigon, *Der Ursprung d. griech. Philosophie*, 1945, have not discussed Clement's testimony at all.

[2] Parmenides frg. B 4, 1, Diels–Kranz, 5th ed., I, 234, 7.

[3] Parmenides frg. B 7, 2, Diels–Kranz, 5th ed., I, 234, 32, *νόημα*.

[4] Parmenides frg. 16, 1 f., Diels–Kranz, 5th ed., I, 244, 8 f. This fragment seems to be closest to the parallel quoted by Clement, cf. *supra* p. 59, n. 4.

[5] This assertion was made, so it appears, by Aristotle, *Metaph.* I.5, 986b, 27 f. = Parmenides frg. A 24, Diels–Kranz, 5th ed., I, 221, 40 f. Some support for it may also be derived from Parmenides frg. B 7, 1 f., cf. *supra*, n. 3, but W. Nestle, *Vom Mythos zum Logos*, 2nd ed., 1942, 111 n. 31, places perhaps too much confidence in it, whilst rightly emphasizing the fact that frg. B 8, 50, Diels–Kranz, 5th ed., I, 239, 6, should not be adduced as proof.

tion of Parmenides' statement, 'the same is to think and to be'. For it has to be remembered that for Parmenides Nous remained capable of being occupied with nonsense as well as with sense, and that it cannot, therefore, be claimed as a principle of true being. Plotinus' mistake was that he substituted for the correct equation of *νοεῖν* = *ἔστιν* an incorrect one maintaining *νοῦς* = *τὸ ὄν*; but it may well be that it served him to express Parmenides' view that the Logos was the 'principle' of reality.[1]

It is the merit of Riezler's book on Parmenides that it insists upon the fact that this philosopher did not simply examine the problems of ontology.[2] This fact can be made evident without much further ado when it is asked why Parmenides insisted upon using *ἔστιν* rather than *τὸ ὄν*, although he did not hesitate to speak of the *μὴ ἐόν*. It is to be regretted that Riezler has nowhere stated this simple linguistic observation, because it would have assisted him in maintaining, as he does, that the problem of Parmenides is rather 'the being of being'. It has to be emphasized that neither the participle *τὸ ὄν* nor even the infinitive *εἶναι* seemed to Parmenides capable of expressing that which he wanted to express. He needed the active *ἔστιν* rather than the participle: Being was to him an activity, even the highest activity. In this there lies a paradox when we see that this conviction is put side by side with his assertion that the cosmos of truth is fettered with chains held by Ananke, necessity.[3] For this allegory is generally understood in the light of the later Eleatic assertion that there cannot be any 'real' motion, as symbolizing that the true cosmos is one of absolute rest. In what way then can this active 'IS' be reconciled with the dominion of Ananke, necessity?

O. Gigon, who seems to have overlooked the paradox just described, has assumed that Ananke was for Parmenides no more than the compelling force of philosophical proof.[4] The fact that this is indeed an important part of her task is well established; but I believe it to be only one

[1] It may be permitted to point out that this result of our research in Parmenides seems very close to John 1:1, all the more since it is assumed that Parmenides was partly dependent upon Heraclitus, C. A. Brandis, *Gesch. d. Entwickelungen d. griech. Philosophie*, 1862, 92 f.

[2] The assertion by O. Gigon, op. cit., 250, 'in Parmenides the conception of being makes its first appearance', although seemingly innocuous, is at least unguarded, and not borne out by either Aristotle or the doxographers.

[3] Parmenides frg. B 8, 26, 30 f., and B 10, 6, Diels–Kranz, 5th ed., I, 235 f., 241, 17 f.

[4] O. Gigon, op. cit., 266 f.

facet of her general, spiritual power. For if the problem of Parmenides was truly 'the being of being', it will have to be regarded as a non-empirical proposition, and Ananke as a metaphysical, indeed eschatological figure, as we find her characterized by Theophrastus.[1] In this light then the world of IS appears as the world of death, a cosmos in which 'of necessity' the only active, and even feasible 'being' permitted consists in reasoning. It is at the same time a world from which there is no return; and it is designed in this way with the intention of opposing the Pythagorean conception of a beyond, which permitted, and even demanded, the migration of the soul from one body to another. But even so this world of IS offered salvation from annihilation, or rather non-existence, the *μὴ ὄν*, the absolute destruction of the human personality in death. Parmenides' doctrine was therefore a kind of gospel for his Greek contemporaries, but a gospel which failed to comfort and inspire the later generations of the Greeks. For men like Aristippus and Epicurus, together with their followers, longed for the relief from the burden of their personality,[2] whilst others, in the mysteries as well as by magical means, strove for their delivery from the despotism of Ananke.[3]

Evidently the world of true being, as outlined by Parmenides, could not be easily reconciled with any conception of *ἀρχή*, and to stress this fact was obviously the intention of Parmenides when he likened the world of IS to a sphere.[4] It has to be admitted, however, that this was an insufficient and indeed somewhat naïve means for achieving such an aim. It seems at least possible, as well as plausible, that at any rate Xenophanes, who in some way was connected with Parmenides, even if the claim made by Aristotle that he had been his master might be

[1] Parmenides frg. A 37, Diels–Kranz, 5th ed., I, 224, 3 f., cf. *supra*, p. 59, n. 1.

[2] Cf. my essay 'Lass die Toten ihre Toten begraben', *Studia Theol.* VI, 1952, 128 f.

[3] An overwhelming wealth of evidence for this may be gathered from the essay 'Die Befreiung Adams aus der Ananke' by E. Peterson, *Frühkirche, Judentum und Gnosis*, 1959, 107 f. In this, as in so many analogous cases, it is to be regretted, however, that the adepts of Hellenistic theocrasy omit to establish any connections with Greek classical thought before proceeding to the evaluation of Near-Eastern analogies. The connection has been established, however, in Grundmann's careful article, *Theol. Wb.* I, 1933, 348, 11 f.

[4] Parmenides, frg. B 8, 42 f., Diels–Kranz, 5th ed., I, 238, 11 f. C. A. Brandis, *Gesch. d. Entwickelungen d. griech. Philosophie*, 1862, 91, concluded that Parmenides described the world of IS as a sphere because he did not conceive of it as bodiless, but it is doubtful whether such an idea needed this special way of expression.

doubted, described the *ἓν τὸ ὂν καὶ τὸ πᾶν* as the supreme *ἀρχή*;[1] and it is certain that Melissus, the pupil of Parmenides, seems to have regarded the true cosmos of Parmenides as the unifying principle of the material world.[2] Parmenides' own description of his world of IS as *ἄναρχον*, i.e. both without beginning and without rule, which I believe to have been dictated by his anti-Pythagoreanism, would not necessarily have excluded such a definition of its relation to the material world. However, it is very unlikely that Parmenides proposed it, although the most part of his philosophy of the world of appearance has been lost to us.

When Parmenides returned from the gate of Dike, the gate of day and night, to this world, he took the third road, the road of the earthly life, which presents to the human mind that mixture of truth and falsehood which Dike had described to him.[3] It is quite clear that only at this point would it have been possible for him to encounter any idea of a 'beginning'. For if the world of IS was *ἄναρχος*, and the *μὴ ὄν* wholly chaotic, it seems logical that the material world, consisting of a mixture of both, should have an *ἀρχή*, a rule or principle. In this respect, however, no direct evidence is forthcoming. We have to rely entirely upon the reports by Aristotle and the doxographers, and their testimony is rather bewildering, far from straightforward, and apparently contradictory. The only advantage over the reports on the Pythagoreans is here that there is no animosity in Aristotle's account of the teaching of Parmenides. In analysing the reports by Aristotle and Theophrastus we may begin with the fact, which has been stated before, that Parmenides' conception of Ananke contained as an important element the idea of the compelling force of logical proof. For it was in this sense that Aristotle presented this conception as the centre of the Parmenidean

[1] Xenophanes frg. A 37, Diels-Kranz, 5th ed., I, 127, 24 f. The apparently cautious wording in which Theophrastus has couched his report, *μίαν δὲ τὴν ἀρχήν, ἤτοι ἓν τὸ ὂν καὶ τὸ πᾶν*, makes me hesitate to draw any far-reaching conclusions from his testimony. It is, on the other hand, the well-known deviation of Melissus from the doctrine of Parmenides, his master, cf. as a preliminary *Evangel. Theol.*, 1960, 576 f., which prevents me from rejecting the reports altogether. For it seems more likely that Melissus returned to an earlier theory, that of Xenophanes, than that he progressed beyond Parmenides. For such a step would appear out of character of a man who had the reputation of being an aristocratic reactionary.

[2] It was Aristotle, *Metaph.* I.5, 986b, 18 f., who combined Xenophanes and Melissus as being *ἀγροικότεροι* than Parmenides, and therefore less worthy of being discussed by him.

[3] Parmenides frg. B 16, Diels-Kranz, 5th ed., I, 244, 8 f. The interpretation of the fragment is disputed.

philosophy. Parmenides, he tells us, 'saw the One under the compulsion of the Logos . . . but the many as existing according to sense perception'. Having stated this, Aristotle continued that Parmenides 'set two causes or principles, hot and cold, appearing as fire and earth. Of these he placed the hot on the side of being, but the other on the side of non-being.'[1] It is here that a certain amount of suspicion is bound to arise. The 'two causes or principles' may have been derived from Parmenides' own words, although the expression exhibits undoubtedly a certain Aristotelian flavour; but it is a puzzle in what way heat and cold can possibly be attached to Parmenides' conceptions of 'being' and 'non-being'. For his conception of IS, as being eternally present, does not tolerate either 'cause' or 'beginning', two ideas which involve a temporal precedence over that which begins or is caused. The 'non-being', on the other hand, may have both in so far as it 'appears' to be; but there the connotation of *ἀρχή*, principle, would render the use of the word inappropriate since the character of the apparent non-being is falsehood, which is essentially un-principled. This being so, it is all the more surprising to see the emphasis laid by Aristotle upon his well-considered duplication: *δύο τὰς αἰτίας καὶ δύο τὰς ἀρχὰς πάλιν τίθησιν*. Our doubts about this report are increased even more when we see Aristotle adding to it that Parmenides was 'compelled (by Ananke) to follow the things which appear'.[2]

It is because of these uncertainties raised by Aristotle's report that the other report on Parmenides, given in the anthology by Theophrastus–Aëtius, demands our attention. It can at least protect us from certain modern conclusions as e.g. that the two expressions meaning necessity, which are to be found in the extant fragments of Parmenides' poem, Ananke and *τό χρεών*, should be treated as synonyms;[3] or that Dike,

[1] Aristotle, *Metaph.* I.5, 986b = Parmenides frg. A 24, Diels–Kranz, 5th ed., I, 221, 40 f.

[2] Ibid., line 30, *ἀναγκαζόμενος ἀκολουθεῖν τοῖς φαινομένοις*. In substance the same report as Aristotle's, although its much cruder form shows signs of having been influenced by Stoic thought, may be found in Hippol., *Philos.* I.11.1 = Parmenides frg. A 23, Diels–Kranz, 5th ed., I, 221, 33 f. Here the careful wording of Aristotle seems to have been spoilt by ignorance, when we read: *οὐδ' αὐτὸς ἐκφεύγων τὴν τῶν πολλῶν δόξαν*. The analogy should yet be evident. Nevertheless the differences between the two testimonies are such that Hippolytus' dependence upon a long school-tradition should be quite evident; and since Aristotle undoubtedly had access to Parmenides' poem, it seems advisable to dismiss Hippolytus as an independent witness.

[3] This seems to be clearly implied by O. Gigon, *Der Ursprung d. griech. Philosophie*, 1945, 253.

sitting at the gate of day and night, does not deserve any more attention than a piece of scenery.[1] For it is certain that if either of these propositions were to be accepted, a just accusation for gross negligence would have to be laid against Theophrastus. Since, however, it seems unlikely that such an accusation could be substantiated, it is safer to use his testimony as a safeguard against such interpretations. In order to achieve this goal, it is however necessary to make a careful comparison between his report and that of Aristotle.

Theophrastus' account of Parmenides' doctrine reads as follows:[2]

Parmenides says that there are wreaths wound the one around another, one of loose texture, the other of dense texture; and others within them, mixed of light and darkness. And the circumference of them all is hard like a wall, beneath which there is a fiery wreath. But of the mixed wreaths the central one is ⟨the origin⟩ and ⟨the cause⟩ of motion and coming into being, which he calls the guiding daemon and the holder of the keys, Dike and Ananke.

When trying to interpret this passage, which gives the impression of not having suffered any change from Theophrastus' original statement under the hands of later redactors, it has to be stated first that the two emendations which have been marked by brackets rest upon the evidence from Simplicius, the Neo-Platonist. He is generally recognized as a reliable witness. These emendations thus appear as a reasonably certain parallel to Aristotle's 'causes and principles'. For it seems improbable that Simplicius, or his authority, transferred them from the *Metaphysics* into the text of Theophrastus, whilst at the same time changing them from the plural into the singular. In this similarity of our two testimonies then there lies the real difficulty for the exegete. It is evident that for Aristotle the plural of 'causes and principles' was essential. For he stressed the fact that there were two: heat and cold; and he aligns them with 'being' and 'non-being', Parmenides' ontological propositions. Theophrastus, on the other hand, only refers to one 'principle

[1] Cf. O. Gigon, op. cit., 246, 'Dike is to be understood, just as in Heraclitus frg. B 94. She watches Helios keeping his course, and arriving and departing at the correct time. She has no further significance.' *Difficile est satiram non scribere.* We may ask whether there is any evidence for Dike being generally regarded as such an heavenly station-master? For the only common feature of the two texts is the name of Dike: In Heraclitus she is 'together with the Furies', and these are absent in Parmenides; whereas Parmenides places her at the gate of night and day, not mentioned by Heraclitus, who rather implies that any deviation from his prescribed course by Helios would receive Dike's immediate attention, and not a reprimand at sunset. It therefore seems unwarranted to connect the two fragments.

[2] Parmenides frg. A 37, Diels–Kranz, 5th ed., I, 224, 3 f., cf. *supra*, p. 62, n.1.

and cause', and avoids any reference to the two ways (or worlds?) of Parmenides. Less difficult may be the solution of a second discrepancy between Aristotle and Theophrastus: Aristotle, it may be observed, makes reference to 'necessity' (Ananke), as does also Theophrastus. However, with Aristotle this is a de-personalized idea, which does not qualify as 'principle and cause'.

Our decision between Aristotle and Theophrastus would be made easy if Theophrastus' report could be rejected on historical grounds, because its author was insufficiently informed about Parmenides. For whilst it would be absurd to assume that Aristotle did not know the entire poem of Parmenides, the same cannot be said about Theophrastus, who lived in the catastrophic age of the wars between the successors of Alexander the Great. However, even from the few lines of the poem of Parmenides which have come down to us we may gain sufficient evidence for the fact that Theophrastus was familiar with it. We find the doctrine of the concentric wreaths and their central daemon in one of the genuine fragments;[1] and we find in another of them that Dike is described as 'the holder of keys', an expression which is also used in Theophrastus' short summary of Parmenides' philosophy.[2]

If, therefore, it cannot be denied that Theophrastus had an independent and first-hand knowledge of Parmenides' poem, it might yet be maintained that his difference with Aristotle does not really touch the doctrine of Parmenides, since the expression 'principle and cause', whether it be used in the singular or in the plural, is plainly Peripatetic. It would thus appear that both Aristotle and Theophrastus aimed at interpreting Parmenides' doctrine rather than at summarizing it. Such an hypothesis would be even correct—up to a point; but even if it were accepted it would still leave us with the question which term used by Parmenides was interpreted by these two Peripatetics as the equivalent for their 'principle and cause', and hardly any other answer can be imagined than: *ἀρχή*. Admittedly the word is not to be found in any of the extant fragments of Parmenides' poem, but it has to be noticed that the word

[1] Parmenides frg. B 12, Diels–Kranz, 5th ed., I, 242 f.

[2] Parmenides frg. B 1, 14, Diels–Kranz, 5th ed., I, 229, 9. The difficulty here is rather one created by Parmenides himself. For nowhere else, so it seems, was Dike entrusted with the power of the keys. The 'key-holder' was normally Hecate, and sometimes also Nemesis–Adrasteia. It is for this reason that I find it hard to believe that Theophrastus too identified the 'holder of the keys' with Dike. It should also be observed that the gates of night and day are not locked, but only bolted, so that they can be opened by Dike's servants without her assistance.

was not only used by Parmenides' own pupil, Melissus, but that Plato also made one of his most consequential statements on ἀρχή in his *Parmenides*.[1] In both these cases the term is used in the singular, and from this fact an argument can be drawn in favour of assuming that in this respect the report of Theophrastus is preferable to that of Aristotle.[2]

This enquiry into the two chief witnesses to Parmenides' doctrine of the 'beginning' had to be made because there is practically no direct evidence of his teaching concerning this world of appearances, the only one of which he could hold that it had a beginning as well as an ending. It remains for us to show how dangerous it would be to form an *a priori* judgment on this point by simply referring to the beliefs of his Eleatic school. For this reason a few words have to be added on the support for our choice of Theophrastus which is to be gained from that pupil of Parmenides, the Samian Melissus. It is well known that this philosopher equated the world of IS in Parmenides' doctrine with the earlier conception of an infinite universe,[3] and thus deprived it of its metaphysical character. He consequently received a sharp rebuke from Aristotle for the somewhat disingenuous way in which he tried to prove the validity of this equation.[4] It is, however, necessary to state that with this equation he also aimed at a proof of the fact that such an infinite universe in its entirety had to be understood as the 'principle' of the world of appearance.[5] I believe that this attempt was made because a mythical 'origin' or 'principle' of the empirical world, similar to that proposed in Parmenides' doctrine, as outlined by Theophrastus, was unacceptable to the 'modern' scepticism of the Samians, amongst whom Melissus not

[1] Plato, *Parmen*,. 245D. Plato here stated plainly that it was the contrast between the two conceptions of ἀρχή and ἄπειρον which had been most strongly emphasized by Parmenides.

[2] The starting point for both Aristotle and Theophrastus can be shown from the extant fragments of Parmenides' poem. It is clear that Theophrastus had in mind frg. B 12, 4 f., Diels–Kranz, 5th ed., I, 243, 3 f.; and Aristotle's report is equally clearly based upon frg. B 19, ibid. I, 245, 15 f. It is this difference which makes the definite choice in favour of one or the other of these two authorities almost impossible, because of the loss of the context of both these sayings.

[3] Melissus frg. B 2, Diels–Kranz, 5th ed., I, 268, 9 f.

[4] Melissus frg. A 10, Diels–Kranz, 5th ed., I, 264, 15 f. = Aristotle, *Sophist. elench.* 5, 167b, 13 f.

[5] Melissus frg. A 13, Diels–Kranz, 5th ed., I, 267, 32 f. The somewhat dubious provenance of this report from an alchemistic treatise *De arte sacra* by the fifth-century Neo-Platonist Olympiodorus, should not deter us from using it at all, as C. A. Brandis, *Geschichte der Entwickelungen d. griech. Philosophie*, 1862, 97 n. 68, seems to suggest, but may entitle us to reject the last two words: τὸ θεῖον.

only enjoyed great political influence, but also strove to introduce his newly acquired philosophy. A mere theory of fire and earth as the 'principles' of the empirical cosmos, in the way in which Aristotle sums up the views held by Parmenides, would not have called for such an ill-favoured attempt at emendation as that of Melissus. He may, therefore, be used, I think, as a witness to the intrinsic correctness of Theophrastus' report.

If such a conclusion seems permissible it may now be stated that Melissus was amongst the earliest Greek philosophers whose interest in 'the beginning' did not centre round its theological aspect. It can neither be held that he gave prominence to any conception of divine rule in the cosmos, nor to its disturbance by human injustice or badness. From what we know of him by the fragments of his work, and the reports of the doxographers about him, his philosophy was a negative one, if perhaps not endowed with the sparkling wit of the later Eleatics. This attitude of Melissus reflected the changed spirit of his time, the middle of the fifth century B.C., when in Ionia and in the Greek homeland voices were to be heard expressing sceptical, agnostic or even plainly atheistic views, of men taking their stand half-way between the theological physics of the Pythagoreans and the wholly untheological physics of the atomists, and in particular of Democritus.

5 Anaxagoras and the atomists

In our pursuit of establishing the meaning of 'the beginning' we have dealt with a very special approach to the problem of the creation of the empirical world in the preceding chapters. We have found that Anaximander threatened it with 'vengeance and retribution in the order of time'. The pythagoreans treated it as a mere accidental to their ideal world of number, and Empedocles had found it ruled by that wickedness from which he had freed himself so that, by escaping from it, he should be deified. Finally, Parmenides, who seems to have been strongly opposed to it emotionally, denied its reality completely, and contrasted it with his non-empirical world of true reality, the world of IS. In the person of Anaxagoras we meet with a much more positive approach to the empirical world. He is reputed to have held that the reason why man should desire to live was 'that he could observe the sky and that order by which the whole world is ruled'.[1] I have searched in vain for a similarly optimistic approach amongst the earlier Ionian natural philosophers; and in this, I feel, there lies the change brought by him to Greek philosophy, whilst his actual doctrines, including his thoughts about the creation of the empirical world,[2] seem to have been closely connected with similar ones held by Anaximenes.[3]

Thus Anaxagoras may be seen in one way as a mere traditionalist; but even here he opened the path for new ideas. For he can be shown to have been deeply conscious of, and at least partly responsible for, the new development of natural philosophy by the introduction of the idea of matter as such, abstract matter. Such an idea was clearly at the bottom of the atomistic philosophy of Democritus, and perhaps already of Leucippus; and it is interesting to observe how Anaxagoras, whilst

[1] Anaxagoras frg. A 30, Diels–Kranz, 5th ed., II, 13, 18 f. = Aristotle, *Eth. Eudem.* I.5, 1216a, 11 f., cf. C. A. Brandis, *Geschichte d. Entwickelungen d. griech. Philosophie*, 1862, 130 n.125. I cannot agree with J. Stenzel, *Kleine Schriften z. griech. Philosophie*, 2nd ed., 1957, saying 'the absence of any idea of guilt is particularly characteristic of Greek thought' (p. 137).

[2] Simplicius, *Comm. in Aristotle, Phys.*, 257b = Anaxagoras frg. A 41, Diels–Kranz, 5th ed., II, 15, 14 f., tries to convince us that Anaxagoras only pretended to hold such a view. Cf. C. A. Brandis, op. cit., 127 n.106, who shows that this assertion is unacceptable.

[3] Simplicius, loc. cit., cf. C. A. Brandis, op. cit., 57 f.

opposing the qualitative equality of the atoms, yet grasped the essential characteristics of 'matter', its tangibility and its divisibility.[1] The advance achieved here may be seen from a comparison with Empedocles, who was only a little older than Anaxagoras, but closely tied to Pythagorean doctrines, as we have seen. Empedocles, therefore, when he approached the problem of the empirical world, made in some way a distinction between his two conflicting 'principles' of love and hatred on the one hand, and the four elements of fire, water, air, and earth, which he acknowledged, on the other. However, it would be rather difficult to outline from the existing sources what exactly this distinction may have been; and the question may well be asked whether he himself ever arrived at a clear delineation of the contrast.[2] Anaxagoras, on the other hand, had a precise idea of the active 'principle', the Nous, which he established and consequently was also fully aware of the fact that that matter which he believed the empirical world to consist of, had to assume a passive character. If this matter was tangible, i.e. determined by its position in space, it did not move of its own in space; if it was divisible, it did not divide itself of its own accord. If the Nous was capable of operating upon matter, such matter would not stretch forth into the domain of the Nous, neither could it be held that the Nous occupied any void since Anaxagoras' doctrine did not admit of any void space. In this respect we notice that Anaxagoras was very close to the opinions of Melissus and the other Eleatics. His was nevertheless a deistic doctrine of the world. Admittedly, he was exiled from Athens on an accusation of atheism,[3] and later generations came to call him 'the atheist',[4] but the notorious atheists of his time, men like Diagoras,[5] however hazy in outline they may appear to us, were men of a very different hue from Anaxagoras.

The deistic approach of Anaxagoras becomes more manifest still when

[1] Überweg–Prächter, *Philos. d. Altertums*, 12th ed., 1926, 99.

[2] Cf. in particular Empedocles frg. A 28, Diels–Kranz, 5th ed., I, 287, 37 f. = Simplicius, *De princ.* XXV.21, a quotation from Theophrastus.

[3] Cf. C. A. Brandis, op. cit., 120 nn.81, 82.

[4] Irenaeus, *Adv. haer.* II.14.2 = Anaxagoras frg. A 28, Diels–Kranz, 5th ed., II, 38, 18.

[5] Cf. about him F. Jacoby, *Abh. d. deutsch. Akad. d. Wiss.*, Mainz, Klasse f. Sprachen etc., 1959, 3. Diels–Kranz, 5th ed., have omitted him altogether; and Überweg–Prächter, op. cit., 129, deny him the quality of philosopher. However, Prof. Jacoby's study has shown the need for a thorough revision of this attitude. If ever a new edition of Überweg's work, which is long overdue, should come to be made, a new appraisal of the fifth-century Sophists so-called will be one of the most urgent improvements.

his philosophy is compared with that of the atomists, Leucippus and Democritus. The difference between them is usually seen in the fact that he did not, as they did, assume that the smallest corpuscules, the molecules or atoms, were without quality, but described them as 'sperms' of the various materials of which the empirical world consists. This difference was caused, however, by the fact that he felt himself bound by the religious—and probably Orphic[1]—conviction of *ἓν τὸ πᾶν*. This fact is clearly expressed in the two sayings traditionally ascribed to him in which the word *ἀρχή* is to be found. These two sayings are mutually complementary. The first of them deals with the nature of matter, and the second with the activity of the Nous.

We begin with Anaxagoras' statement about matter. For it is the introduction of this new idea of 'something' of which created objects must consist, which has rendered the conception of creation so questionable, especially since it was not always realized—and certainly not by Anaxagoras—that the step from 'creation' to 'creation from' is a *transitio in aliud genus*. Anaxagoras stated what follows:[2]

And since the parts are equal (in number?)[3] in size of the great objects as of the small, this also shows that in each thing all things are contained. Neither can there be any separate existence,[4] but everything has its share in everything.[5] Since there

---

[1] This view is based upon O. Kern, *Orph. frg.*, 1922, 158, frg. 165, 1. However, the verse does not establish it beyond doubt that already in the fifth century B.C. the Orphics knew the poem in which it appears now. For this poem is found only in two Neo-Platonists of the beginning of the sixth century A.D., in Proclus. *Comm. in Tim.* (twice), and in Olympiodorus, *Comm. in Phaed.* What is certain, however, is the fact that from Xenophanes onwards, Parmenides and the Eleatics, on the one hand, and Empedocles on the other, held the *ἓν τὸ πᾶν* as a religious conviction, cf. the references in Diels–Kranz, 5th ed., III, 339, 24 f.

[2] Anaxagoras frg. B 6, Diels–Kranz, 5th ed., II, 35, 13 f. The various points where I differ from Kranz's translation will be marked in the subsequent notes.

[3] W. Kranz translates *ἴσαι* as 'gleichviel', but that seems begging the question. For normally a large object may be divided into a greater number of parts than a small one. I assume, therefore, that Anaxagoras here referred to the equality of his different 'sperms' with regard to their size. For *πλῆθος* in this sense cf. Liddell–Scott s.v. III.1.

[4] This is Kranz's translation. Another possibility would be the translation, 'neither is being anything apart', with a polemical edge against Parmenides; but such a translation might be regarded as an over-interpretation. However, Kranz's translation treats this statement as no more than a repetition of the preceding clause, and such a treatment does not seem very satisfactory either.

[5] In this clause the ambiguity of *μοῖρα*, meaning both 'share' and 'destiny' has to be stressed. For it refers evidently to the Greek conception of the cosmos as a 'body', cf. my remarks in *Ztschr. d. Savigny Stift.*, rom. Abt., 1953, 305 f.

is no possibility for anything to be the smallest, no smallest object could either be separated or come into being of itself;[1] but as it existed simultaneously (with all the others) from the beginning, so does the universe in the present exist simultaneously. In all things, however, there is a multifariousness of substance[2] and they are equal in size (proportion?) in the objects which have been segregated (sc. from original matter) whether they be big or small.

The second passage,[3] which is the most substantial of Anaxagoras' sayings that have come down to us in the original Greek, and which describes the active principle of the Nous and its action, cannot be retailed in its full length here. In it the philosopher, having first of all rehearsed once more his doctrine of *ἓν τὸ πᾶν*, describes the Nous as the finest of all substances, which is not capable of entering into a union of material objects unless they belong to that exceptional group of those which have a soul. Subsequently, continuing his description of the Nous, he states:[4]

And also of the entire cyclical movement the Nous has taken charge and control,[5] so that it (the universe) turns round from the beginning (*τὴν ἀρχήν*).

Before discussing the contribution to the conception of 'the beginning', and its change brought about by Anaxagoras' doctrine, we shall consider the use of these two passages in such a connection. For we find in both of them that the word *ἀρχή* is used adverbially; and it appears from W. Kranz's index to the Pre-Socratics, as well as from his translation of our two fragments, that he was not inclined to put any great store by the use of the word in its present context. Without entering into the grammar of Anaxagoras' fragments, it may be held that this approach is not to be recommended on account of the general trend of his philosophy. We have seen already that Anaxagoras maintained that the empirical world had been 'created'; and the cyclical motion of its constituent parts is not proposed as signifying its eternity. Consequently, since in the idea of creation the element of a 'beginning' cannot be omitted, the word for it, wherever and in whatever grammatical

[1] W. Kranz translates *γενέσθαι* as 'to be', which may appear as a post-classical usage.

[2] Added by W. Kranz.

[3] Anaxagoras frg. B 12, Diels–Kranz, 5th ed., II, 37, 18 f.

[4] Diels–Kranz, 5th ed., II, 38, 5 f.

[5] *'Εκράτησεν*. W. Kranz's translation 'hat der Geist die Herrschaft angetreten' rightly emphasizes the element of action, which is expressed by Anaxagoras' use of the aorist.

construction it appears, may justly claim that special attention should be given to it.

Some further remarks of a general character may be made in support of this claim: Anaxagoras, although he maintained that the empirical world was created, nevertheless held that it was the 'real' world. He thus avoided the dichotomy of the Pythagoreans, and of Parmenides,[1] who had claimed 'true reality' only for their ideal worlds of number, and of IS, and had granted to the empirical world no more than a derived reality, if that much. The realism of Anaxagoras appeared clearly when he not only proclaimed that all the constituents of the empirical world were 'material' in essence, but even demanded that this world's moving power, the Nous, was also material. It has been argued that this was only done by him because Greek philosophy at this time found it difficult to describe an immaterial existence.[2] However, such an argument is at least inconclusive, if not actually incorrect, since it presupposes that there was an *a priori* need for defining the substance of the Nous which, I feel certain, cannot be claimed. The material character of Nous was stated by Anaxagoras on the assumption that that which is material is also real. As regards the substance of the empirical world, there seems to me to have been an antinomy in Anaxagoras' thought. For he claimed, on the one hand, that it was created from his 'molecules'—the least particle of gold still being gold, that of bone still being bone etc.[3]—and on the other that it was infinite 'from the beginning'. For it seems logically impossible to combine a 'beginning' with infinity,[4] unless this infinity as such is regarded as 'the beginning', in which case we are thrown back upon the philosophy of Anaximander, and are no longer free to accept Anaxagoras' different conclusions. Such was indeed the opinion adopted by Theophrastus,[5] but Anaxagoras seems not to have felt thus committed, and in fact such a conclusion seems never to have been drawn.

There is one special feature of Anaxagoras' philosophy which deserves

[1] I can find no trace of any 'système des deux plans', ascribed by J. Zafiropulo, *Anaxagore de Clazomène*, 1948, 298, to the philosophy of Anaxagoras, and I venture to remain sceptical with regard to Zafiropulo's psychological *a priori* considerations, by which he attempts to prove its existence.

[2] Cf. C. A. Brandis, *Geschichte der Entwickelungen d. griech. Philosophie*, 1862, 124.

[3] Cf. Lucret., *De rer. nat.* I.835 f. = Anaxagoras frg. A 44, Diels–Kranz, 5th ed., II, 17, 29 f.

[4] Cf. Aristotle, *Phys.* III.5, 205b, I = Anaxagoras frg. A 50, Diels–Kranz, 5th ed., II, 20, 9 f., and its interpretation by Brandis, op. cit., 127.

[5] Cf. Anaxagoras frg. A 41, Diels–Kranz, 5th ed., II, 15, 33 f.

to be stressed because it may account for the influence subsequently exercised by it particularly upon the Stoics, and to a lesser degree also upon the Christians. The friend of Pericles and victim of that arch-reactionary, Thucydides the son of Melesias, although he agreed with Melissus that 'everything' was potentially contained 'in the beginning',[1] he attempted to construct a 'democratic' philosophy, in contrast to the other philosophers whom we have discussed so far. For Anaxagoras based his philosophy upon the axiom that it was the many (minute molecules) which in themselves represented and together constituted the whole, and not the whole from which each individual thing received its being.[2] Once his three presuppositions, (*a*) that the empirical world is real, (*b*) that its principal parts, whether they are active or passive, share in the same reality of substance or matter, and (*c*) that the 'equal particles', as Aristotle calls them, on the one hand form a universe which allows of no void, and on the other are destined to conglomerate in specific objects—I say once these three basic tenets of Anaxagoras' cosmogony are clearly grasped, the significance of the 'beginning' for his philosophy will be plain, and the adverbial use of *ἀρχήν* will appear as accidental. It will be seen that in such a system the 'beginning' was bound to have the character of rulership and government rather than of principle. The question to be answered is, however, whether Anaxagoras considered only the Nous, the active partner, as *ἀρχή* or whether he granted the totality of the equal particles the same qualification. If, as I believe, the latter was the case, the assertion made,[3] that action rather than order was the characteristic of Anaxagoras' system, may stand in need of some rectification.

It is undoubtedly true that Anaxagoras in his philosophy placed the Nous as the supreme ruler. This appears most clearly from certain say-

[1] Melissus frg. B 11, Diels–Kranz, 5th ed., I, 276, 2 f., cf. *supra* for its authenticity, and Anaxagoras frg. B 4 fin., Diels–Kranz, ibid., II, 35, 3 f.

[2] Cf. my *Politische Metaphysik* I, 1959, 111 f. The mode of the 'original state' of matter and Nous, according to Anaxagoras, are still under discussion, cf. W. Spoerri, *Späthellenistische Berichte etc.*, 1959, 14 n., cf. ibid., 15, the noteworthy remark about the difference between him and Democritus. It may well be an over-interpretation by Aristotle that Anaxagoras ever considered an 'existing' original matter. I think that he posited it only as a 'beginning'. The assertion by W. Nestle, *Vom Mythos zum Logos*, 2nd ed., 1942, 182 f., that Anaxagoras' 'theoretical life' was intentionally 'non-political' seems unconvincing, cf. on the part played by the political debate in the development of Pre-Socratic philosophy K. v. Fritz, *Studium Generale* XIV, 1961, 605 f. Anaxagoras was a 'professional' philosopher as much as Pericles was a professional politician, and these are democratic conceptions.

[3] Cf. W. Nestle, op. cit., 185.

ings of Plato's who, although he showed himself in his *Phaedo* the most devastating critic of Anaxagoras' philosophy,[1] was yet deeply influenced by it. It is, therefore, not surprising to find that Plato strongly supported the result which we have reached here by *a priori* considerations, and said:[2]

The just exists, of which Anaxagoras holds that it is the Nous; for he calls it the absolute ruler (αὐτοκράτωρ), and mixing with nothing else, but ordering all things by pervading them all.

The quotation, for such it seems to be, nevertheless leaves us in doubt whether or not it had been Anaxagoras himself who had personified the Nous by calling it 'the absolute ruler'. For in the *Statesman*, a dialogue which is admitted by all to be under considerable obligation to Anaxagorean thought, it is Socrates himself (and that means Plato) who describes the ruler of the cosmos as 'the absolute ruler', if in a somewhat ambiguous sense. For his absolute rulership is derived from 'the helmsman', the controlling deity, and its absolutism is shown by the misuse of his power.[3] In contrast to this Platonic presentation, it seems significant that one of the few intact fragments of his work which are still in existence,[4] and which incidentally seems to have been Plato's source, preferred to use the neuter αὐτοκρατές. The change of gender in Plato gives us the possibility to maintain that the description of the Nous as ἀρχή in Anaxagoras' philosophy was not interfered with by mythological speculation.[5] Aristotle accordingly, when summing up Anaxagoras' doctrine of the Nous, wrote as follows:[6]

[1] Plato, *Phaedo*, 97C f., cf. J. Stenzel, *Kleine Schriften z. griech. Philosophie*, 2nd ed., 1957, 130, 142 f.

[2] Plato, *Cratyl.*, 413C. Here too Plato's—or Socrates'—report makes it clear that he rejects Anaxagoras' view.

[3] Plato, *Polit.*, 274A, cf. ibid., 298C, 299C, where the word is clearly used, though not by Socrates himself, in a pejorative sense, meaning 'arbitrarily'. The same meaning is also to be found in *Legg.* IX.875B, whereas ibid., IV,713C, it is Chronos who is 'the absolute (and just) ruler'.

[4] Anaxagoras frg. B 12, Diels–Kranz, 5th ed., II, 37, 19.

[5] The claim made by W. Nestle, *Vom Mythos zum Logos*, 2nd ed., 1942, 185, 'this unbounded and autocratic spirit, the finest and purest substance existing, omniscient and omnipotent, is nothing else than the deity', may perhaps have some validity as a modern appreciation of Anaxagoras' thought, but seems wholly alien to his philosophical intention.

[6] Aristotle, *De anima* I.2, 405a, 15 f. = Anaxagoras frg. A 55, Diels–Kranz, 5th ed. II, 20, 30 f.

He places the Nous above all things *as the beginning*. For he says this alone of all things is simple and unmixed and pure. He also attributes to the same beginning the power of both to know and to move, saying that the Nous is moving the universe.

It has been claimed that 'rule by the Nous consists in its having power to move the molecules at will, but in a downright mechanical manner'.[1] If this were correct Plato's polemic in his *Phaedo*, and his description of the Nous as 'just' would be a blatant misrepresentation of Anaxagoras' doctrine. However, if J. Stenzel was right in using the support of Diogenes of Apollonia in his interpretation of the conflict,[2] and I think that he was, it would be unsafe to deny the ethical character of Anaxagoras' conception of the Nous for the simple reason that it works by mechanical means. This recognition in its turn shows that Aristotle too was right when he accorded an ethical character to Anaxagoras' conception of the Nous as ἀρχή: 'Anaxagoras says that goodness is the moving power *from the beginning*. For the Nous is the mover, but it moves for a purpose.'[3]

If then it has to be assumed that according to Anaxagoras the Nous acted independently and, as Socrates complained, in the likeness of the great leaders of nations, the establishment of order amongst the multitude was its inherent task. It is therefore necessary to cast a glance at this multitude. Aristotle's report that it consisted of 'homoiomeres', as he—and perhaps Anaxagoras himself—called the molecules, has been subjected not so long ago to a very severe castigation,[4] according to which it should be regarded as an invention arising from his superficiality and ill-will towards Anaxagoras. It has been maintained that Anaxagoras' doctrine of the unlimited divisibility of matter logically excluded all atomistic theories. This thesis is hardly convincing. For if Anaxagoras was serious when he held that 'in itself everything is simultaneously great and small',[5] then his doctrine of an unlimited divisibility did not involve aim at an unlimited reduction of quantity.[6] It rather denied

[1] F. M. Cleve, *The philosophy of Anaxagoras*, 1949, 22. The term used there, 'at will', is ambiguous. However, in view of the fact that Cleve has italicized the '*mechanical* manner', I assume that he denies the Nous the power of ethical decision.

[2] J. Stenzel, *Kleine Schriften*, 2nd ed., 1957, 142.

[3] Aristotle, *Metaph.* XII.10, 1075b, 8, quoted after F. Trendelenburg, *Aristotelis De anima*, 2nd ed., 1877, 193 f.

[4] J. Zafiropulo, *Anaxagore de Clazomène*, 1948, 276 f.

[5] Anaxagoras frg. B 3, Diels–Kranz, 5th ed., II, 33, 17 f.

[6] Cf. J. Stenzel, op. cit., 140 f., 'Parmenides' idea of being is insolubly connected with the mathematical concept of magnitude, the form of the world in space;

reality to the standard of quantity. If, therefore, quantity was excluded in this context from the rank of philosophical category, an infinity of molecules may be presupposed in Anaxagoras' system so long as it is understood that he distinguished between the concepts of division, which is possible, and separation, which is not. This proposition in its turn, however, limits the activity of the Nous. For it demands that the multitude must be regarded as a unit, and can only exist as a unit, i.e. can only be acted upon as a whole by the Nous.

The Nous then is the active principle, which however is only qualified to do just this, but may not separate anything from the total. Aristotle has put this conviction of Anaxagoras in a form which shows clearly the philosopher's awareness and partial use of contemporary Pythagorean conceptions:[1] 'He named as *ἀρχαί* the One—for this is simple and un-mixed—and as the other that which we set down as the indefinite, before it is defined and assumes any kind of quality (*εἴδους τινός*).' The 'One' is obviously Aristotle's interpretation of the Nous; but it may yet appear as a fairer approach than that of Plato in that Aristotle abandons the personification of the Nous as an 'autocrat'. On the other hand, it appears that Aristotle has lost sight of the ethical moment in Anaxagoras' conception of the Nous, and has thus turned his philosophy into a somewhat inadequate physical speculation. The same has to be said with regard to the reports of the doxographers upon whom we have to depend very largely when we now examine the interaction between the Nous and the 'indefinite'. Here we have first to beware of a pitfall: Theophrastus reports that 'Anaxagoras was the first to change the doctrines concerning the *ἀρχαί*.'[2] I believe that this statement is true if the meaning of 'origins' or 'principles' is adopted for its translation. At the same time it is most probable that Theophrastus

---

Anaxagoras, however, seems to have made all the terms of the Eleatics to refer to quality'; F. M. Cleve, *The philosophy of Anaxagoras*, 1949, 12.

[1] Aristotle, *Metaph.* I.8, 898b, 16 f., Diels–Kranz, 5th ed., II, 21, 20 f. The editors have joined this saying to *Metaph.* XII.2, 1069b, to form Anaxagoras frg. A 61. This procedure may be open to criticism. For the sentence quoted here comes from the well-reasoned critical appraisal of Anaxagoras' out-moded doctrine of matter in *Metaph.* I.8, 898a, 30–b.20; but the first half of frg. A 61 belongs to that obscure passage in *Metaph.* XII.2, which since long has defied the ingenuity of exegetes, cf. W. Spoerri, *Späthellenistische Berichte etc.*, 1959, 14 n.18. The formation of frg. A 61 has not, to my mind, solved the puzzle of *Metaph.* XII.2. The same contrast of the One and the indefinite is claimed for Anaxagoras once more by Aristotle, *Phys.* I.4, 187a, 20 f. = Empedocles frg. A 46, Diels–Kranz, 5th ed. I, 291 f.

[2] Anaxagoras frg. A 41, Diels–Kranz, 5th ed., II, 15, 14 f.

meant by the word 'the elements', and in this sense the statement is misleading. Admittedly, this latter meaning was very common in the third century; but in the fifth it was not. Theophrastus, therefore, may give us here an insight into Anaxagoras' theory of matter, but presumably not into his theory of 'the beginning'.

More important is the fact that Theophrastus described 'matter' as conceived by Anaxagoras as the 'passive cause', *αἰτία ἐλλείπουσα*. We hear from Simplicius[1] that he summed up his findings about Anaxagoras' doctrine in the following manner:

> It seems that Anaxagoras makes the material *ἀρχαί* (the elements) infinite in number, but the causes of motion and coming into being only one, the Nous. If therefore it is assumed that the mixture of them all is the one physis, indeterminate as regards quality (*εἶδος*) and quantity (*μέγεθος*), it follows that he states two principles (*ἀρχαί*), the physis of indetermination, and the Nous.

This summary by Theophrastus is admirable from the point of view of abstract logic, but perhaps less so from that of historical accuracy. For it is first of all certain that Anaxagoras never thought of a logical contrast between 'the physis of indetermination' on the one hand, and the Nous on the other. The Nous, it is true, was not supposed to amalgamate with any other matter; nevertheless it was a 'physical principle', not 'spirit', which, as J. Stenzel has rightly stated,[2] was immanent in the universe of those constituents of which it formed the cosmos.

Although the Nous, according to Anaxagoras, was a physical principle, it was at the same time, if we may believe Plato, a moral principle, and in both these capacities it was the creator of the cosmos. This creation, which was effected by setting the molecules into a cyclical motion, commenced of necessity in one minute point,[3] and spread from there amongst the constituents of the previously unmoved 'indeterminate physis' of the *πάντα ὁμοῦ*, which potentially contained everything, though not perceivable through sense perception. A similar conception we have found already in Melissus, and the close affinity between the thoughts of Anaxagoras and the Eleatics is indeed a well-established fact. However, the part played by Nous in the system of Parmenides is,

[1] Ibid., Diels–Kranz, 5th ed., II, 15, 30 f.

[2] J. Stenzel, *Kleine Schriften z. griech. Philosophie*, 2nd ed., 1957, 142, has based this conclusion upon certain fragments from Diogenes of Apollonia, Anaxagoras' contemporary. I can find no reason why this evidence for the interpretation of Anaxagoras should not be used.

[3] Cf. the references given by J. Zafiropulo, *Anaxagore de Clazomène*, 1948, 305.

as we have seen, an ambiguous one, whereas for Anaxagoras it is an omniscient power. Thus a recent writer on Anaxagoras has stated:[1]

Nous is a being that works consciously. And so is the omnipotent God of the Bible who creates the world out of nothingness to be subservient to His ends.—But there is a difference. Nous, not being a Creator, is only cognizant of what will result from his interferences.

We may take notice of this interesting comparison, but have to qualify the expression 'interference', which seems to imply a casualness in the work of Nous which, although it may have a place in Eleatic philosophy, seems alien to Anaxagoras' doctrine. With this caveat, it will be our last task in dealing with Anaxagoras' theory of 'the beginning' to examine which of its particular features may have conjured up such an association.[2]

The fact is that it was Anaxagoras who was perhaps the first to set the scene for a doctrine of the Nous as the ultimate cause, but at the same time regarded it also as the supreme principle of the world. In short, it was his aim to combine in his conception of 'the beginning' both the normative and the mechanical element, claiming its identity whether it was considered externally or from the inside of the empirical cosmos. It seems nevertheless true to say that in this combination the Nous as the normative principle was of much less importance for Anaxagoras than the Nous as the first mover. For he had found that an explanation of the empirical world had to concentrate upon its functioning rather than upon its being. The fact that Anaxagoras in doing so rather neglected the ontological approach to the supreme active 'beginning' brought about not only the passionate protests of Socrates and Plato, but also the frankly materialistic attempt of his pupil Archelaus.[3]

[1] F. M. Cleve, *The philosophy of Anaxagoras*, 1949, 26.

[2] With regard to the claim that Nous was no creator, reference has to be made to Anaxagoras frg. B 17, Diels–Kranz, 5th ed., II, 40 f., where he maintains that in the empirical world there is no coming into being and no annihilation, but a cyclic change of the mixture of the elements, cf. F. M. Cleve, op. cit., 12.

[3] Archelaus frgg. A 4, 5, Diels–Kranz, 5th ed., II, 46, 4; 26 f., drawn from Hippolytus and Simplicius, show that Theophrastus in the case of this philosopher too described the molecules as *ἀρχαί*, but it may be noted that there was no original matter combining them all to form an ultimate passive 'beginning' as in Anaxagoras. It is rather asserted that for Archelaus heat and cold were *ἀρχαί*, frgg. A 4, 8, Diels–Kranz, op. cit., II, 46, 5 f.; 47, 3 f., i.e. that these two were no longer regarded by him as 'matter' in the Anaxagorean sense. And it was in the molecule that Archelaus placed the ontological question, cf. Plato, *Soph.*, 242CD, saying that he called them *τὰ ὄντα*. That makes it impossible to determine the part which was

From him one may gain an illustration of the influence which the Eleatic discussion of the problem of motion exercised upon the philosophy of his master Anaxagoras. For it is evident that in the Greek philosophy of the fifth century B.C. the effect of any cause, including even the first cause, was seen as being only one and the same, namely motion. In particular, change of quality, if it was allowed at all as being real—Anaxagoras would have questioned this—was subsumed at this early time under the general concept of motion;[1] and it seems to me that the Eleatic denial of the reality of motion will not be fully understood if it is not regarded as containing a protest directed against the uncritical acceptance of the thought-form of cause and effect.

Whilst stating this inevitable result of a logical, *a priori* deduction it has to be admitted that there seems to be only little evidence that its validity was recognized and discussed by subsequent philosophers. It is, however, permissible I believe to quote a passage from Plato's *Laws* in support of the view proposed, in which a hidden reference to Anaxagoras is made.[2] At the same time it is necessary to inspect the Eleatic argument against the reality of motion, and the reasons for its refutation, because it is of vital importance for the character of that empirical cosmos for which 'the beginning' was found to consist in an ultimate cause. For a 'beginning' which was characterized as the supreme principle did not necessarily call for 'time' as a constituent of the empirical world. It was not illogical to do so. We have seen Chronos playing a

---

played by the Nous in the philosophy of Archelaus, cf. frg. A 10, Diels–Kranz, op. cit., II, 47, 9 f., Augustine's report on inanimate molecules. Still less credible is Epiphanius' report, frg. A 11, that Archelaus regarded the earth as *ἀρχή*; but frgg. A 6, 7; 12, and perhaps 14, drawn from Sext. Emp., that he held that the Nous was air, have the support of Diogenes of Apollonia, cf. J. Stenzel, *Kleine Schriften z. griech. Philosophie*, 2nd ed., 1957, 142, and are probably correct.

[1] This is clearly shown by the fact that Socrates in Plato, *Theaet.*, 181B f., described change of quality as 'the second type of motion'. That this was traditional is indicated by the fact that in *Legg.* X.894C, Plato, having outlined previously (893B f.) eight different kinds of motion, adds two more to make them up into the 'perfect' number of ten. On the metaphysical significance of motion in Greek thought cf. J. Stenzel, op. cit., 16 n.43.

[2] Plato, *Legg.* X.895A: *εἰ σταίη πως τὰ πάντα ὁμοῦ γενόμενα . . . τίν' ἄρα ἐν αὐτοῖς ἀνάγκη πρώτην κίνησιν γενέσθαι*; cf. J. Stenzel, op. cit., 17. Here, it seems to me, Plato criticized the casual character which Anaxagoras appeared to have given to the beginning of motion by the Nous, always assuming that the *πάντα ὁμοῦ* had a static existence. But was this Anaxagoras' view? Plato's reference, ibid., to the *πλείστοις τῶν τοιούτων* makes me hesitate, cf. the hidden reference to Anaxagoras in *Legg.* XII, 967B.

part in Anaximander's philosophy; but it seems to be significant that even the word 'time' is absent from all fragments and testimonies of Parmenides' philosophy.[1] If, on the other hand, 'the beginning' was seen as the supreme cause, time (and space) were necessary to the empirical cosmos on the basis of the syllogism of *post hoc—propter hoc.*

The possibility or impossibility of motion was proposed by Zeno of Elea[2] in his famous problem of 'the arrow'. Aristotle, who once more is our earliest and most important witness, has treated the problem from two points of view when he attempted to solve it. He reports that Zeno had said, 'whether the universe is in constant rest or motion, as long as it is in accord with itself a thing in flight is for ever in the present, and thus the flying arrow does not move'. Aristotle accordingly first tackled the problem without reference to the general motion or immobility of the universe, and resolved it by reference to the continuity of time and motion. Afterwards he discussed it again from the point of view of a moving universe, examining the nature of separate motion within a moving system. It seems, however, that in his two short remarks he somewhat too quickly passed over the suggestion made by Zeno, and perhaps directed against Empedocles, that separate motion, as in the motion of the flying arrow, constitutes a disturbance within the harmoniously constructed—or functioning—universe. For this argument cannot be fully resolved by reference to the continuity of time and motion, but demands a further consideration of the problem of motion within the system of the cosmos. The reason for this is that it states that a moving object, the arrow, if moving within the cosmos, passes through a space which is filled with air or some other medium, which in its turn demands to be taken into account. The critic of Zeno, therefore, has to distinguish between two possibilities. If the arrow flies through

[1] Cf. W. Kranz's index, Diels–Kranz, 5th ed., III, 475 f., s.v.

[2] Zeno frg. A 29, Diels–Kranz, 5th ed., I, 253, 38 f. = Aristotle, *Phys.* V.9, 299b, 5 f., 30 f. Aristotle, cf. H. Rüdiger, *Der Kampf mit dem gesunden Menschenverstand*, 1938, 41 f., has treated this, and indeed all Zeno's problems, as a purely logical one; but it may be asked whether such a treatment is really satisfactory. For it neglects not only the metaphysical side of the problem of motion, which had been so strongly stressed by Plato, cf. *supra*, p. 80, n.1, but also its physical side. For it may be asked e.g. whether it is really true that the 10,000th part of a millet falling to the ground produces the one-millionth part of the sound produced by 100 millets falling the same distance on the same spot so that Zeno's proposition is really incorrect, seeing that sound is produced in a medium like air etc., which would prevent such a minute particle from falling with the same speed as that of the bigger —and heavier—body.

the air, it may be held that harmony of the cosmos is being maintained by the fact that this medium is capable of a balancing counter-movement, filling the space left void behind the flying arrow at the same speed at which the arrow leaves it. This, I take it, is what Anaxagoras meant when he described the Nous as starting a gyratory movement; but it may be open to the question whether or not such balancing counter-movement does not cancel out the initial movement of the arrow so that in reality no movement has taken place. By the same token such an answer leads of necessity to the distinction between fluid matter, within which movement may take place, and solid matter, like steel or stone, which does not allow of movement taking place within it because it is not capable of a balancing counter-movement. The second possibility is to assume that the empirical world is porous, full of holes, into which the matter opposing the arrow can escape. In this case the relation between the flying arrow and the opposing matter which escapes from its course could not be determined by any conception of balancing volumes, but only by that of the density of the opposing matter on the one hand, and by the energy of the motion of the arrow on the other. The volume of displaced matter might be greater or smaller than that of the arrow, and when that view was taken it was fair to argue that, unless the density of matter was held to be equal everywhere, all motion resulted in disturbing the harmony of the cosmos, or to put it more bluntly, such a universe if it admitted of any motion was no cosmos, no harmonious system at all, and no rational pronouncement could be made about it. From this necessary corollary to Aristotle's 'solution' of the problem of 'the arrow' it follows that the doctrine of the impenetrability of matter, which was to be developed later by Plato, was already common to the Eleatics and to Anaxagoras, whose 'material' Nous, though immanent in the empirical world by its function, had no standing, no being, within it.

It was this conviction which radically distinguished Anaxagoras' philosophy from that of the atomists, particularly Democritus. For it was he who chose the second answer; and it appears from his terminology that he did so in direct and intentional contrast to Parmenides and the Eleatics.[1] For he described the void by which, as he contended, action was made possible, as the *μὴ ὄν*, that which Parmenides had passionately

[1] J. Stenzel, *Kleine Schriften z. griech. Philosophie*, 2nd ed., 1957, 61, has claimed that 'historically and materially atomism should be seen as a consequent development of Eleatism which, however, advanced beyond it in certain respects'. I hesitate to accept this verdict.

denounced as having no reality, and consequently put out of bounds for all philosophical enquiries. The doxographers accordingly claimed for Democritus that his philosophical system rested upon two ἀρχαί, the void and the atoms.[1] These doxographical reports, however, show every sign of being Hellenistic evaluations of Democritus' atomistic theory rather than straightforward historical accounts. Wherever the word ἀρχή appears in them it is used in a loose sense as 'data' without either a normative or a causative connotation; and it seems safe to maintain that such a use of the word was quite uncommon amongst the writers of the fifth century B.C. if it is to be found at all in their writings. It is also possible, so it seems, to show a certain development of this usage. For in Simplicius we find that he first recorded that in Democritus' cosmology the ἀρχαί were 'the void' (τὸ κενόν) and 'the full' (τὸ πλῆρες), but subsequently interpreted these two terms as 'the non-being' and 'the being'.[2] The opposite is to be found, however, in that silly Christian apology *The ridiculing of the philosophers* by an otherwise unknown writer Hermeias, and it seems to me that in this form it may go back to the original report by Theophrastus.[3] For in this form the pair may have appeared to the pupil of Aristotle as two 'principles' forming a contradictory contrast. However, if such was the case, Theophrastus himself was in this instance guilty of misrepresenting Democritus. For we learn from Aristotle[4] that the philosopher from Abdera did indeed comprehend the aggregate of all the atoms under the heading of the 'common body', despite their varieties in size and form; and that this was done in order to characterize this conglomeration as ἀρχή. He said nothing, however, of the 'void' having been characterized in the same way; and it would probably have been an infringement of the thesis that 'nothing arises out of nothing', if Democritus had done so, a thesis to which Democritus himself adhered, and which Aristotle was prepared to defend strongly.

It is, nevertheless, true that Democritus felt the necessity of arranging the 'full' and the 'void' into a pair of logical opposites.[5] This can be

[1] Democrit. frg. A 1, Diels–Kranz, 5th ed., II, 84, 10 f. = Diog. Laert. X.4.

[2] Democrit. frg. A 38, Diels–Kranz, 5th ed., II, 94 4 f.

[3] Simplicius, loc. cit., n. 53. Hermeias, *Irrisio* 12 = Democrit. frg. A 44, Diels–Kranz, 5th ed., II, 95, 20 f. H. Diels seems to have believed that Simplicius quoted verbatim from Theophrastus–Aëtius, but I cannot believe that Hermeias had the wit to change the sequence.

[4] Democrit. frg. A 41, Diels–Kranz, 5th ed., 91, 1 f. = Aristotle, *Phys.* III.4, 203a, 33 f.

[5] Cf. Democrit. frg. B 125, Diels–Kranz, 5th ed., II, 168, 5 f. Logically the μὴ ὄν

shown by his discussion of the primary and secondary qualities of perceptible objects. Secondary qualities, which are entirely dependent upon the frequently deceptive perceptions of the senses, and therefore for the most part, if not altogether, have no reality, are colour, taste, temperature, etc., and the atoms have none of them.[1] It is, however, incorrect to claim that the atoms have no qualities at all, as is often done. For they possess the primary qualities of volume, shape, and perhaps gravity.[2] And these qualities are shared by them with the observable objects constituted from them and the 'void'. There is moreover one attribute which is peculiar to the atoms, which caused Democritus to describe them by the unconventional term of *τὰ ναστά*,[3] the attribute of solidity. This theory shows an evident opposition to the cosmology of Anaxagoras at least in two respects: First, that the atoms have such a very limited number of qualities, whilst Anaxagoras claimed for his minute corpuscules that they comprehended all qualities within themselves; and secondly, that the atoms have no origin,[4] whereas Anaxagoras, as we have seen, postulated an active and a passive beginning, the Nous, and the *πάντα ὁμοῦ*. In this respect we have even some direct evidence for Democritus' attack upon Anaxagoras. For Aristotle[5] reports his saying, 'one and the same are the active and the passive'. This, Aristotle assures us, was Democritus' new recognition, and had not been derived from any earlier philosopher. It is, therefore, safe to assume that it was meant to counter Anaxagoras' doctrine of supreme causes, and perhaps the entire conception of causality. Christian critics in particular accused Democritus of ascribing all that happened to Fortune and necessity.[6]

Aristotle, on the other hand, gave unstinted praise to Democritus

---

cannot, of course, be opposed to 'the existing' since this would demand that 'nothing' should be comparable to 'something'. J. Stenzel, *Kleine Schriften zur griech. Philosophie*, 2nd ed., 1957, 62 f., has shown that Plato's *Sophist* was conceived in order to oppose Democritus' atomism by establishing this very point.

[1] Democrit. frg. B 125, *supra*, p. 83, n. 5, cf. frgg. A 49; B 11, Diels-Kranz, 5th ed., II, 97, 10 f.; 140, 9 f.

[2] But cf. Democrit. frg. A 47, Diels-Kranz, 5th ed., II, 96, 1 f., where Aëtius claims that only Epicurus added gravity to this list.

[3] Democrit. frg. A 46, Diels-Kranz, 5th ed., II, 95, 27 f.

[4] Cf. Democrit. frg. A 56, Diels-Kranz, 5th ed., II, 98, 31 f. = Cicero, *De fin.* I.17, *eumque motum atomorum nullo a principio, sed ex aeterno tempore intelligi convenire.*

[5] Aristotle, *De gen. et corr.* I.7, 323b, 10 = Democrit. frg. A 63, Diels-Kranz, 5th ed., II, 100, 24.

[6] So Dionysius of Alexandria and Lactantius, cf. Democrit. frgg. B 118 and A 70, Diels-Kranz, 5th ed.

for being the first to introduce the scientific method, by which he meant his own method of investigation, i.e. the inductive method.[1] This praise may have been well deserved, but it should not blind us with regard to the precariousness of the fundamentals upon which the atomistic theory, from Leucippus to Epicurus, established its conclusions. For it was soon remarked how arbitrary it was to admit the impenetrability of the smallest particles within the universe, and yet to deny it to any larger, observable objects.[2] Democritus had to counter this criticism by maintaining that a compact conglomeration of any multitude of atoms was made impossible by their difference in size and shape;[3] but apparently he did not realize that in this way he jeopardized the most important assumption of his atomistic theory. For maintaining, as he did, that these qualities belonged eternally to the atoms,[4] he transferred their resistance against a fusion with other atoms from their 'solidity' to their surface, and thus deprived this their supreme quality of its meaning, since the resistance of the surface made an inspection of the substance of the atoms logically impossible. On the other hand, shape and size being geometrical conceptions, he opened his atomistic theory to a Pythagorean interpretation; and we have already seen that the Pythagorean, Ecphantus, made use of this opportunity.

Secondly, it was strongly maintained by Democritus that 'the nothing exists as much as the thing'.[5] This saying was bound to involve the philosopher in the discussion about the idea of *creatio ex nihilo*, which at this time was seriously proposed by certain Sophists. For if the empirical world—or any empirical world, for Democritus claimed the existence of an unlimited number of cosmoi—had for its constituents the equally important 'thing' and 'nothing', and if it was subject to coming into being and dissolution in time, it came into being, at least partly, 'out of nothing'. It seems more probable, therefore, that this was the point at issue between Democritus and Xeniades of Corinth, one of those

[1] Democrit. frg. A 36, Diels–Kranz, 5th ed., II, 93, 14 f.

[2] J. Stenzel, *Kleine Schriften z. griech. Philosophie*, 2nd ed., 1957, 61, stating, 'wherever we read about division an ultimate indivisible is logically demanded', is unconvincing. It is true that the process of division will never reach 'nothing'; but it is also true that it will never reach an ultimate indivisible.

[3] Democrit. frg. A 37, Diels–Kranz, 5th ed., II, 93, 29 f.

[4] Cf. in particular Democrit. frg. B 141, Diels–Kranz, 5th ed., II, 170, 3 f., where it is said that Democritus treated *ἰδέα* meaning 'shape', as synonymous with *τὸ ἐλάχιστον σῶμα*.

[5] Democrit. frg. B 156, Diels–Kranz, 5th ed., II, 174, 18, *μὴ μᾶλλον τὸ δὲν ἢ τὸ μηδὲν εἶναι*, from Plutarch, *Adv. Colot.* 4, 1108F, cf. frg. A 57, ibid., p. 98 f., originating from the same source.

Sophists, than the theory of recognition based by Xeniades upon his cosmology.[1] For, as we have shown elsewhere,[2] Democritus' scepticism was hardly less outspoken than that of his opponent; but he could not admit the theory proposed by Xeniades that the only constituent of the empirical world was the non-existent. It is, however instructive to see how anxiously Diogenes Laertius in his summary of Democritus' philosophy avoids the mention of Democritus' equation between 'the void' and the non-existent, in order to protect him from any suspicions of bordering on Xeniades' revolutionary theory.[3]

The dispute between Democritus and Xeniades is of great significance because it opens up a view on the main purpose of Democritus' philosophy. Democritus is widely hailed as one of the great founders of the scientific method, and he rightly deserves this praise. However, an inspection of the sources shows him to have been—like Aristotle—as great a moralist as he was a scientist. The scientific method was for him not an end in itself, but a means to establish the truth. This was possible only, he held, by establishing the phenomena of the empirical world as real. In this quest he proved himself a master of critical scepticism, who had to reject the unlimited scepticism of Xeniades. The empirical world was not all sham; it was a mixture of the 'solid' reality of the atoms, and the equally necessary irreality of 'the void', nothingness. When we view it in this light we find that the problem of 'the beginning' was attacked by the three immediate predecessors of Plato in three different ways. Parmenides and the Eleatics regarded it as identical with the problem of being, the ontological problem. Anaxagoras saw it as a problem of functioning, and consequently developed the theory of causality. Finally, Democritus treated it as the problem of reality and truth, and found an answer to it in that theory, which he regarded as a hard fact, of the 'solid' atom. It seems as if what we have in Plato is a synthesis of the propositions of the Anaxagoras school (that of Democritus being accepted the least enthusiastically and that of the Eleatics the most enthusiastically) with the Pythagorean proposition of mathematical order in his vision of the cosmos rather than a complete system of cosmology.

[1] The quotation from Sext. Emp., *Adv. math.* VII.53, ed. I. Bekker, 1842, in Democrit. frg. B 163, Diels–Kranz, 5th ed., II, 176, 15 f., seems to me too short, because it omits the crucial *καὶ ἐκ τοῦ μὴ ὄντος πᾶν τὸ γινόμενον γίνεσθαι*, and W. Kranz's 'version': 'mention of (and objection to) X.'s theory of recognition', which may claim an uncertain support from Sext. Emp., *Adv. Math.* VIII.5 f., ed. I. Bekker, p. 289, 26 f., but hardly from the text he quotes.

[2] A. Ehrhardt, 'Creatio ex nihilo', *Studia Theologica* IV, Uppsala, 1950, 23 n.1.

[3] Diog. Laert. ix, 34–49.

6 Plato

Plato, to whose cosmology we now turn, nowhere mentioned in any of his writings the name of Democritus.[1] I assume that this was largely due to the fact that—with the one exception of the 'battle of the giants' in his *Sophist*—he was anxious to avoid polemics, almost at any price. Anaxagoras, as we have seen, and Parmenides he mentioned because of their important contribution to the philosophy of his teacher, Socrates; but otherwise even the name of Pythagoras occurs only once in his work, in the tenth book of his *Laws*. It is, therefore, difficult to assess the influence of these his predecessors upon his philosophical views, and remarks concerning such influences may often remain hypothetical. It is still more difficult to make any final statement about Plato's own philosophical views. For his work shows that his own immense mind was constantly developing, even within the very writing of one or the other dialogues of his; and the amount of work spent since then upon the elucidation of his thought is so great, and so weighty, that it seems beyond human powers even to read the bulk of modern research, let alone scrutinize it. All that is intended here is to reach an understanding of his conception of ἀρχή, mainly on the basis of two of the dialogues of his middle period: the *Parmenides*, and the *Phaedrus*, written after his first journey to Sicily, and of parts of two of his latest period, the *Timaeus*, and the tenth book of his *Laws*. Even within these limitations I have had to entrust myself to a more experienced pilot: J. B. Skemp's book on Plato's theory of motion[2] has been an invaluable guide, and whilst on occasion I have had the temerity to disagree with it, I sincerely hope that my sense of admiration and gratitude for the book has never deserted me.

The choice of Skemp's book for a guide is due to the fact that Plato's theory of the 'beginning' is consistently a theory of a 'beginning of motion'. Here also lies the deepest root of certain disagreements between my views and those of Professor Skemp. For while he refers repeatedly to 'the Ionian conception of ἀρχή', as if it were an established basis upon which Plato could found his theory of motion, and which he finally could use for his doctrine of the soul, and of the demiurge,[3] our

[1] Cf. J. Stenzel, *Kleine Schriften z. griech. Philosophie*, 2nd ed., 1957, 60 f.

[2] J. B. Skemp, *The theory of motion in Plato's later dialogues*, Cambridge, 1942.

[3] J. B. Skemp, op. cit., cf. especially the remarks on page 24, dealing with Plato, *Polit.*, 269D f.

enquiries have not led to the conclusion that there was any established 'Ionian conception of ἀρχή'. All that may be said to have been common in the use of the word amongst the Pre-Socratics who employed it, is that it marked the problem how, if at all, the finite could spring from the infinite or, inversely, the infinite could ever touch the finite. This character of 'mark' or 'sign' was clearly shown even when Democritus described the atoms as the ἀρχαί of all the empirical, finite cosmoi, the existence of which he maintained. It cannot be held however, that before Anaxagoras 'the beginning' was understood as the beginning of motion. We have even seen how desperately Zeno of Elea tried to defend the position, taken by his master, Parmenides, that there was no motion, but presumably tolerated without criticizing it, the use of ἀρχή in Parmenides' poem.

In Plato's *Parmenides* the first great question which concerns the subject of our enquiry is the question of the relation between the divine, perfect knowledge of the true reality of the world of ideas, and the imperfect human recognition of this world, based upon the sense perception of the empirical world. The deity, being perfect, cannot know anything imperfect, including the imperfect philosophical considerations of the human mind.[1] This proposition contains once more the chief question of all theories of creation, the question how the infinite can ever come into contact with the finite. The answer attempted by the Platonic Parmenides is to be found in his subsequent discussion of the One, which is his conception of the deity. He stresses, right from the outset of his teaching, that the One has 'no beginning, middle, or ending'.[2] In the progress of the discussion on the One it is further stated by Parmenides that 'beginning, middle, and ending' in time are as meaningless when they are applied to the One as their counterparts in space. For the One is not only infinite in space, but also eternal in time.[3] Moreover, the One is incapable of establishing any connection with the dyad on its own initiative; and yet such a connection takes place.[4] In addition a whole number of contradictory contrasts are established both in the thesis that the One is real, and that reality is one.[5] Finally, it is contended that even the apparently existing smallest bodies (ὄγκοι), which are to be found in that which is other than One, are equally without 'begin-

[1] Plato, *Parmenides*, 134A f., esp. 134C.

[2] Ibid., 137D, cf. the identical completion of the argument, 145A f.

[3] Ibid., 153B f.

[4] Ibid., 149D, τὸ ἓν τῶν τε ἄλλων καὶ ἑαυτοῦ ἅπτεταί τε καὶ οὐχ ἅπτεται.

[5] Cf. the useful and instructive tables in C. Ritter, *Plato* II, 1923, 70 f.

ning, middle, and ending'.[1] It appears, therefore, that the dialogue in which ἀρχή in the technical meaning of the word is most frequently used by Plato, puts up a very definite warning against any incautious use of the word. It contains four out of the eight cases in which the word is technically used by Plato, and in all these cases its philosophical applicability in the sense of a first or original beginning is denied.

The second part of the *Parmenides* is said to constitute a puzzle, and it appears as if its ending is inconclusive. One impression, however, which the reader can hardly escape, is that it succeeds in showing the contradictions inherent in the popular saying ἕν τὸ πᾶν; and once it is seen that Zeno, the companion of Parmenides, used the phrase prominently in his work,[2] it seems likely that in this part a vindication of the young Socrates was attempted. For at the end of the first part of the dialogue Parmenides, having summed up the imperfections of the argument of the young Athenian, also criticizes the way in which Zeno had proposed his thesis that a 'multitude' of objects exists. Zeno, Parmenides maintains, had omitted the counter-check about the existence of the One;[3] and if, as I assume, Zeno's treatise was actually known to Plato and his contemporaries, the final discussion about the minute particles may well have demolished its findings. At the same time it has to be noticed that it is not just 'the beginning', but 'the beginning, the middle, and the ending' which are repeatedly denied any reality. This formula, we remember, was not of philosophical origin but had its habitat in the religious, theosophical or mythological poetry of the Orphics.[4] It may, therefore, be considered hypothetically that the Parmenides of Plato's dialogue was actually Plato himself, represented as proclaiming a purer religion, or at least a purer theology than Orphism —and presumably also Pythagoreanism—at least in their revivalist form. On the other hand, it seems evident from the treatment of the ὄγκοι, the smallest particles, that Plato's *Parmenides* also meant to show that taking plurality for a starting point, as Democritus was doing, could not provide a more convincing, in fact not even an essentially

[1] Plato, *Parmenides*, 165A f.

[2] Cf. Diels–Kranz, 5th ed., III, 149a, 10 f.

[3] Plato, *Parmenides*, 136A f.

[4] Cf. O. Kern, *Orphic. Fragm.*, 1922, 90 f., 201, frgg. 21, 168, and elsewhere. It seems to me that this observation may be of some significance for assessing the distance between Plato's *Parmenides* on the one hand, and his *Phaedrus* and *Timaeus* on the other. It is not by chance that the *Parmenides* contains no myth, but because its whole tendency is critical and destructive rather than constructive.

different, solution for the problem of the reality of the empirical world.[1] More doubtful is the question whether the repeated assertion that the One in general, and in particular the non-existent One, cannot have a 'beginning', was also directed against Democritus who, as we have seen, cannot be shown to have described the *μὴ ὄν* as an *ἀρχή*.

If Plato's *Parmenides* erected a warning against the Orphic use of the formula 'beginning, middle, and ending', at least in the philosophical discussion, it is true to say that he gave in his *Phaedrus*, a dialogue which was probably written only a short while before the *Parmenides*,[2] a far more positive account of the meaning of *ἀρχή*. Here he made the first attempt of combining it with the ideas of motion and of the soul. Before analysing his argument, two general observations must be recorded. The first that this idea seems to have occupied the mind of Plato only since his return from his first journey to Sicily;[3] the second that during this period Plato exhibited a certain diffidence in discussing this problem, which involves the relations between 'the beginning' and the conception of motion in all its forms. This becomes particularly apparent in the *Phaedrus*, where 'the beginning', and its connection with the soul and with motion, are discussed in the third speech of Socrates, leading up to the *Phaedrus*-myth, a speech ascribed by the speaker to Stesichorus, son of Euphemus, from Himera in Sicily,[4] presumably the famous lyrical poet. It seems probable that Plato tried in this way to excuse himself of some of the responsibility for the opinions expressed:[5]

Every soul is immortal. For that which is ever-moving is immortal. But the other (objects), which move as well as are moved from elsewhere, as soon as their motion comes to an end, their lives also come to an end. Only the self-moving, since it cannot abandon itself, never ceaseth to be in motion; but at the same time it is also to all those other objects which it moves, the source and the 'beginning' of motion. But the 'beginning' itself is without origin. For out of the

---

[1] U. v. Wilamowitz–Möllendorf, *Plato* I, 1919, 558, has emphasized that Plato used Parmenides in the *Sophist* as the anti-materialistic protagonist, and this seems equally true already for the *Parmenides*.

[2] Cf. C. Ritter, *Plato* I, 1910, 265, but see ibid., 263 n.1, where the possibility is admitted that the *Parmenides* and the *Phaedrus* were contemporary. He gives cogent reasons why *Phaedrus* should not be regarded as one of the earliest Platonic dialogues.

[3] The word *κίνησις* is found once in *Cratyl.*, 426C, but the discussion of its significance belongs entirely to our group of dialogues, *Phaedrus*, *Theaetetus*, *Parmenides*, and *Sophist*, on the one hand, and *Timaeus* and *Laws*, on the other.

[4] *Phaedrus*, 244A.

[5] Plato, *Phaedrus*, 245C f.

'beginning' all that has come into being takes of necessity its origin; but ('the beginning') itself out of nothing. For if the 'beginning' took its origin out of anything (else), it could not take it from the 'beginning'. And since it has no origin, it is also necessary that it does not perish. For if the 'beginning' should perish it could neither come into being out of something, nor could anything out of it come into being, if so be that everything must come into being out of the 'beginning'. Thus the 'beginning' of motion is that same (being) which moves itself. And it is impossible for it either to come into being or to perish; for otherwise the whole heaven and all creation would fold up, and come to a stop, and never have any chance of coming into being again by being moved.

Before discussing this passage we must emphasize once more the difference between the conception of *ἀρχή* here, in the *Phaedrus*, from that which Plato rejected in the *Parmenides*. There 'beginning, middle, and ending' were seen as ideas which of necessity belonged together, in time as well as in space. At the same time they were seen as having an existence of their own if they had an existence at all. In *Phaedrus*, however, the 'beginning' is taken separately, but not independently: it is the beginning of motion, if equated with 'life'. This appears as the one fixed point throughout the whole argument, which otherwise is full of ambiguities.[1]

The passage quoted, and at least its continuation to p. 246A, but perhaps even the entire third speech of Socrates, is, as Professor Skemp has rightly remarked, only loosely connected with the preceding dialogue, and as far as it has been quoted not even very significant for the subsequent myth.[2] It begins quite clearly as a confession of faith: 'Every soul is immortal.' This assertion, the *thema probandum*, is then put to the test. The proof is based upon two theses. The first is that life is motion; and the second is that all motion springs from one ultimate source, the self-moving mover. At first sight these two assertions have a compelling

[1] J. Stenzel, *Kleine Schriften z. griech. Philosophie*, 2nd ed., 1957, supported Natorp's thesis that the chief difficulty arises on p. 246C, in the distinction between mortal and immortal beings, both represented as joining a soul to a body, either for a time or for everlasting. This dilemma, however, is no more than a consequence of the use of the plural 'every soul' at the beginning of the passage quoted here.

[2] J. B. Skemp, *The theory of motion*, 1942, 3. I do not think that Professor Skemp himself is of the opinion that already in this passage the combination of 'the Orphic-Pythagorean ideas of the soul' with 'the Ionian conception or *ἀρχή*' was fully achieved. F. Schleiermacher, *Platons Werke*, 2nd ed., 1.1, 1817, 379 f., commenting on our passage, had the same impression. He, wrongly ascribing the *Phaedrus* to a very early date in Plato's production, explained this impression by assuming that Plato, prior to his journey to Sicily, could only have a superficial knowledge of Pythagoreanism. If this is an injustice to Plato it might yet be true for Stesichorus.

force; but closer inspection makes it clear that neither of them is really convincing. For if it is accepted that without motion there can be no life, it is still not evident that where there is motion there is also life;[1] and thus some stronger reason seems to be required to support the first thesis. The second thesis is also controversial. For we have seen in the case of Anaximander that the early Ionian natural philosophers had already treated hylozoism as a serious proposition; and it may also be held that this doctrine had been revived, if only in a specific sense, by the atomistic theory of Democritus, even if it is true that he distinguished between living motion and purely mechanical motion.[2] In that case then the primary, fortuitous 'whirl' of the atoms, and not a 'self-moving mover', would still figure as the ultimate source of both types of motion. Both the propositions put forward to support the immortality of the soul are, therefore, controversial, and Plato—or Stesichoros?—is aware of this when he says:[3] 'This proof will be unconvincing to the rationalists,[4] but reliable to the wise.'

Such a hint at the existence of an esoteric doctrine must not be overheard; it does not, however, exonerate us from enquiring into the philosophical antecedents of Plato's Creed. Professor Skemp has suggested that our passage was dependent upon 'the Ionian conception of ἀρχή, and related to the 'flux principle of Heraclitus'. We have already expressed our reserve with regard to the first ingredient, and feel slightly uneasy also with regard to the second.[5] However, we need not take up here this argument with Professor Skemp in view of the fact that he himself admits that in any case the conception of ἀρχή in the *Phaedrus* is far removed from Heraclitean thought. Here he suggests that Anaxagoras was the source for Plato's conception. Following Burnet, he tries by means of a very subtle enquiry to disprove Aristotle's allegation that Anaxagoras had seen the Nous as the unmoved mover.[6] I am afraid

[1] From the parallels in the *Laws* and the *Timaeus*, quoted by J. Stenzel, op. cit., 3, it seems clear that Plato himself held this view, in so far as the astral bodies were concerned, already in *Phaedrus*, 246C.

[2] Cf. Cl. Bauemker, *Die Theorie der Materie*, 1890, 65.

[3] *Phaedrus*, 245C, ἡ δὲ ἀπόδειξις ἔσται δεινοῖς μὲν ἄπιστος, σοφοῖς δὲ πιστή.

[4] F. Schleiermacher's translation.

[5] It seems to me that such Heraclitean influence as there may be, was mediated by the Heraclitean school rather than by the master himself. The slight admixture of acerbity which is noticeable in Professor Skemp's (*The theory of motion*, 1943, 32 n.3) criticism of K. Reinhardt, *Parmenides*, 1916, 291, is partly, I think, caused by the fact that an enquiry into Plato's obligation to Heraclitus was then, as it is still, a desideratum.

[6] J. B. Skemp, op. cit., 33 f.

that he has not succeeded in making me his follower in this respect. Quite apart from a desire of concentrating upon Sicilian,[1] i.e. Pythagorean precursors, I see Anaxagoras' doctrine in a different light. For in Anaxagoras' system the Nous appears almost as an equivalent to 'the void' in Democritus' atomistic cosmology. For only because Democritus rejected the hypothesis of an existing active, rational cause, was he compelled to reduce the Nous to non-existence, and to ascribe to the atoms the ability of producing life. On the other hand, it is difficult to see a moving material Nous in an Anaxagorean cosmos completely filled with 'matter'. For being matter itself, such a Nous could hardly escape that mixture with ordinary matter of which it is allegedly incapable. Whichever view may be correct, Professor Skemp's or ours, the acceptance of a plea for Anaxagorean influence upon our passage in the *Phaedrus* would still demand an answer to the awkward question, in what way the Nous of Anaxagoras was identified with that Psyche which, as the 'self-moving mover', obtained the place of the 'beginning' in the speech of Stesichorus in the *Phaedrus*. Such then are the reasons why the theory promoted by Schleiermacher[2]—by instinct rather than by argument—that the speech of Stesichorus in its totality rests upon views proclaimed in the mysteries, seems the most promising approach to its understanding.

Before it is attempted to enquire into the relations between the doctrine of the mysteries and Plato's doctrine of the 'self-moving mover', it is necessary to account for the apparent contradictions between the approach to the problem of ἀρχή in his *Parmenides* and in his *Phaedrus*.[3] For it seems yet possible to shed some light upon the *reservatio mentalis* with which young Socrates is made to accept the staggering, and not always well-founded, assertions of 'father Parmenides'. Such a reserve on Plato's part seems to be indicated by the fact that in the first part of

[1] Because of its Sicilian origin I would attach some importance to a saying of Epicharmus Ennii frg. B 49, Diels–Kranz, 5th ed., I, 206, 8 f., *eius (sc. agriculturae) principia sunt eadem quae mundi esse Ennius scribit, aqua, terra, anima et sol.* The fragment shows that this Pythagorean poet-philosopher from Sicily, Plato's contemporary and friend, counted the soul amongst the *principia mundi*, and not the Nous.

[2] Cf. e.g. F. Schleiermacher, *Platons Werke*, I.I, 2nd ed., 1817, 57.

[3] This paragraph is meant as an attempt at progressing beyond the somewhat resigned summary of *Parmenides* by J. B. Skemp, op. cit., 14, where the solution of the paradoxes contained in this dialogue is said to arise from certain hints in Plato's *Philebus* and *Sophist*. This seems largely true with regard to the theory of motion, which is Professor Skemp's primary concern. There remains, however, *ein Erdenrest zu tragen peinlich* with regard to ἀρχή, which these two dialogues discuss only by implication.

the *Parmenides* the problem of motion is only mentioned in passing;[1] and that its discussion takes place in the second part only, which has generally nihilistic tendencies, and is not specially addressed to Socrates.[2] Some elucidation of the relation between that 'beginning, middle and ending' which is denied by Parmenides,[3] and the other self-moving, infinite and eternal 'beginning' of motion, as mentioned in the *Phaedrus*, may be derived from Aristotle's discussion of 'ideal' motion. For Aristotle contends that the 'ideal' motion is circular motion. This he contrasts with straight motion, which he classes as an inferior type of motion. Several reasons are offered for this classification, the convincing power of which varies. One of these reasons, the last, has a direct bearing upon our question:[4]

> In rectilinear motion we have a definite starting point, finishing point, and middle point. . . . On the other hand, in circular motion there are no such definite points. For why should any one point on the line be a limit rather than the other?

This argument is used in order to show that only circular motion can be continuous, and therefore eternal. For rectilinear motion is of necessity interrupted motion since continuous motion along an infinite straight line is logically impossible. All this makes good sense mathematically; but the emphasis laid upon the superior dignity of the potentially eternal circular motion reminds us of the mystic connotation of the formula 'beginning, middle, and ending'. For the question of 'dignity' has no place in the field of mathematical reason.

This observation alone, however, does not help sufficiently because in the dialogue Parmenides expressly excluded the circular movement of the One,[5] together with the 'beginning, middle, and ending' of it.[6] Plato, so it seems, was not greatly perturbed when abandoning these two; but it is here where the argument from silence may count. For since a difference in the dignity of the various forms of movement has appeared, it is worth our while to look out for a still worthier movement. Here it

[1] Plato, *Parmenides*, 129E.

[2] Ibid., 138B, 156B f.

[3] Another witness to the fact that this denial belonged to the 'real' Parmenides, may be derived from Metrodorus of Chius frg. A 4, Diels–Kranz, 5th ed. II, 231, 21 f., who also mentions his 'mark' of immobility.

[4] Aristotle, *Phys.*, 265a, quoted from S. Samburski, *The physical world of the Greeks*, 1958, 87.

[5] Plato, *Parmenides*, 138C f.

[6] Later, when discussing the ideal proportion, Plato avoided the use of *ἀρχή*, and spoke, *Tim.*, 31CD, of *πρῶτον*, *μέσον*, *ἔσχατον*.

is noteworthy that the gyratory movement of the spindle has been omitted by Plato from the consideration of his Parmenides. Yet this form of movement is shown, at the end of the myth of the *Republic*, to be 'the beginning' of human life,[1] and most scholars now consider the tenth book of the *Republic* as contemporary with the *Parmenides*. Such, therefore, was the motion of the soul, the self-moving mover, being a beginning rather than having a beginning, yet not absolute motion, but controlled motion. For the self-moving mover was not just a *perpetuum mobile*; it was divine, was the soul.

If in this way a possibility has appeared that Plato as the author of the *Parmenides* was not at variance with himself as the author of the *Phaedrus*, it can now be attempted to draw a line from the *Phaedrus* to the *Timaeus*. In an admirable and convincing analysis of the evidence provided by the doxographers, Professor Skemp has shown the close connection between the doctrine of the soul held by the Pythagorean philosopher and physician Alcmaeon of Croton and by Plato in his *Timaeus*.[2] Whilst that much is admitted it seems that the 'warning' added by Aristotle as a rider to Alcmaeon's thesis, 'men perish because they are unable to join the beginning to the end', may have a significance different from that suggested by Professor Skemp. Aristotle added to Alcmaeon's saying: 'This[3] is clearly said if one accepts it is a generalization of his, and makes no effort to relate the saying to details.' I cannot help feeling that this may not have been an attack upon Alcmaeon so much as upon Plato himself. J. Stenzel rightly drew attention to the unsolved contradiction between the fates of the *ζῷον θνητόν*, on the one hand, and the *ζῷον ἀθάνατον*, on the other, in the *Phaedrus*; and he rightly asked the question whether the human soul before entering or after leaving its human body does not appear as purer than the divine soul, which is for ever joined to its body, 'joining the beginning to the end'.[4] If it should be objected that there is no sufficient evidence for

[1] Plato, *Rep*. x, 620c.

[2] J. B. Skemp, *The theory of motion*, 1942, 36 f., esp. 39 f. Professor Skemp (p. 37) regards Alcmaeon as a contemporary rather than a pupil of Plato, a view which is probably correct. In that case they must have met, probably in Sicily, for it cannot be denied that the fragments of Alcmaeon's work reflect in many respects the opinions of his greater contemporary.

[3] Aristotle, *Problem*. XVII.3, 916a, 33, quoted from J. B. Skemp, op. cit., 40 no. v. Alcmaeon's view is also reflected, I think, in Plato, *Tim*. 36BC, 41B.

[4] J. Stenzel, *Kleine Schriften z. griech. Philosophie*, 2nd ed., 1957, 2 f., cf. esp. note 4, discussing *Phaedrus*, 246C. J. Stenzel does not mention Alcmaeon and Aristotle, but it seems unlikely that Aristotle should have been unaware of this serious flaw in the Phaedrus myth. F. A. Trendelenburg, *Aristotelis De anima*, 2nd ed., 1877, 196,

giving Aristotle's attack such a specific character, and I admit that the evidence is not conclusive, Aristotle's remark should still be treated as an altogether negative criticism, meaning: 'This is a clever camouflage of an esoteric doctrine.' I cannot believe, as Professor Skemp does, that Aristotle 'accepted the formula, but rejected the underlying doctrine'. At least I know of no evidence supporting the claim that he accepted the formula. On the other hand, it has to be said that Professor Skemp has established convincingly the close connection between Alcmaeon's formula and the Phaedrus myth. The 'beginning' in the sense of the Phaedrus myth or, at least, of the passage quoted above, is here seen very largely as the concern of the individual soul. There seem to be as many 'beginnings' as there are souls.

It has been shown by J. Stenzel that Plato's *Timaeus* and *Laws* can be used for the interpretation of the Phaedrus myth without the risk of importing alien matter.[1] It is, therefore, a promising task to examine the use of *ἀρχή* in these two writings so as to find whether such a theory of innumerable beginnings, so to speak of spiritual molecules as opposed to Anaxagoras' material molecules, was indeed held by Plato.[2] The first use of the word occurs right at the beginning of the speech of Timaeus:[3]

The whole heaven, therefore, or the cosmos . . . with regard to it it has to be found out first, as the basis of everything else, whether it has always been, without a 'beginning' of its coming into being, or whether it came into being, taking its origin from some 'beginning'. It has come into being. For it is visible and tangible, and has a body; for all suchlike objects are open to sense perception. And all these things, which can be open to a theory based upon observation, are evidently liable to becoming and to being created.[4]

---

commenting on *De anima* I.2.17, refers to the witness in *Metaph.* I.5, 986a, 26, where it is said that the eternity of its motion proved to Alcmaeon the immortality of the soul, cf. Diog. Laert. VIII.5. Since Aristotle does not refer to the Phaedrus myth anywhere in his *De anima*, it seems tempting to assume that this myth was derived from Alcmaeon's doctrine, and for this reason did not demand a separate treatment.

[1] J. Stenzel, op. cit., 3–6.

[2] Earlier theories about the soul, as in *Phaedo*, 79C–80B, cf. E. Rohde, *Psyche* II, 5/6th ed., 1910, 269 n.5, were still at work here, and in *Rep.* X, 611A, with its emphasis upon the eternity of all the numerous souls. E. Rohde, op. cit., 278 n.2, discussing this passage, omits completely the creation of souls, *Tim.*, 41D, and J. Stenzel, op. cit., 7, analysing the *Timaeus*, omits to mention *Rep.* X.

[3] Plato, *Tim.* I, 28B.

[4] The analogy to Paul, 2 Cor. 4:18, 'for the things which are seen are temporal etc.', is evident. There are more Platonic parallels quoted by E. Spiess, *Logos Spermatikos*, 1871, 297, all of them similar to, but none identical with, the Pauline saying.

This passage clearly bears the mark of its Pythagorean origin in the words, 'the heaven or cosmos'. Elsewhere I have shown that this equation was the original Pythagorean concept;[1] and in its light the Phaedrus myth can also be shown to have been drawn from the Pythagorean well,[2] but not the passage dealing with ἀρχή, where 'every soul' (πᾶσα ψυχή) without distinction is characterized as being 'the self-moving mover'.

The progress made by Plato in his *Timaeus* is shown clearly by the change of his starting point. He no longer put the question of the 'beginning' from the point of view of the individual human soul, not even from that of the eternal soul in general, but from that of the general harmony of the entire cosmic system. That becomes abundantly clear when the second passage in the *Timaeus* in which he deals with ἀρχή is analysed. For in this passage he introduces the 'world-soul' which, as we hear in a later remark in the speech of Timaeus,[3] is markedly superior to all the individual, and especially the human, souls. On this 'world-soul' Timaeus informs his audience:[4]

> But this [i.e. the 'world-soul'], woven through from the centre towards heaven everywhere and enwrapping it [sc. the empirical cosmos], and itself turning round within itself, has begun a divine 'beginning' of an unending and intelligent life throughout all time. And indeed, the visible body of heaven has come into being, but the soul, being invisible as well as partaking in reason (λογισμός) and harmony to the eternal things of the mind (τὰ νοητά), has been created by him who is the best, as the best of all things created.

In this passage it is clearly stated that the eternal motion of the cosmic soul is gyratory, a form of motion which we have postulated already for the proper appreciation of the second part of Plato's *Parmenides*. We also learn from it that the cosmic soul penetrates through the whole of the cosmos, meaning the empirical, material world—for not being matter itself, it is capable of doing that which the Nous of Anaxagoras was unable to do—but also envelops it; and finally that this cosmic soul

[1] Cf. *Politische Metaphysik* I, 1959, 149 n.3.

[2] J. Stenzel, op. cit., 7 f., who also stresses the basic identity of the views expressed in the Phaedrus myth and in the speech of Timaeus, calls them 'Orphic'; but it seems to me that the Pythagorean elaboration of these 'Orphic' opinions should not be overlooked.

[3] *Tim.*, 41D, where the creation of these individual souls is described as 'mixing them in an approximately equal fashion (sc. as the cosmic soul), yet no longer equally pure, but as a second or third (edition)'.

[4] *Tim.*, 36E–37A.

is the 'beginning' (*ἀρχή*) of the empirical world in the sense that it mediates between the world of ideas, the things of the eternal Nous, and creation.[1] For although the cosmic soul itself is created in the sense that it owes its being to the father and demiurge, this has to be seen as an event which has constituted the cosmic soul within the world of ideas as it were by right, whereas the empirical cosmos received in its creation a perpetual motion by sufferance only.[2] This short summary is meant to make it clear that Plato here aimed higher than at the vindication, and at the same time determination, of the claims of Parmenides, and perhaps other members of the Eleatic school. His lucid and penetrating mind was indeed fully aware of the problem posed by Anaximander, who had introduced the new philosophical conception of *ἀρχή* in order to signify the inescapable demand for a mediation between the infinite and the finite.

In order to set this first 'beginning' into relief, reference has now to be made to the existence of a second 'beginning' in this 'best possible world',[3] the 'beginning' of evil. In so far as the evil in this world is connected by Plato with the pollution of the soul by the body,[4] it does not touch the subject of our enquiry. However, there existed the 'unspeakable' in the empirical world.[5] Timaeus had to explain that the good (*καλός*) demiurge was not omnipotent. The eternal father had been obliged to persuade, with the help of the Nous, the dark power of Ananke to grant him permission that he might arrange the heaven as the best possible cosmos. Such leave was necessary. For we are told that there is another *ἀρχή* or *ἀρχαί*, of which Timaeus is unwilling to speak, although he gives us to understand that it is not only the elements.[6]

[1] Cf. the exposition by J. Moreau, *L'âme du monde*, 1939, 4–6.

[2] Cf. the distinction made between *génération* and *fabrication*, J. Moreau, op. cit., 7.

[3] Cf. *Tim.*, 29A. J. Moreau, op. cit., 8, sums up Plato's prominent intention as follows: 'L'univers, qui comprend la totalité du donné . . . est donc nécessairement la plus belle des réalisations, et l'activité d'où elle procède de toutes la plus parfaite.'

[4] The significant passages from Plato have been carefully collected and discussed by E. Rohde, *Psyche*, 5/6th ed., 1910, II, 270 f.

[5] *Tim.* 29A, *εἰ δέ, ὃ μηδ' εἰπεῖν τινι θέμις, πρὸς τὸ γεγονός*, does not mean that there was only the possibility of the demiurge not being honest, but that by taking the empirical world as real, and copying it, the power of evil had entered the world.

[6] *Tim.*, 48A–D, where the special prayer said by Timaeus illustrates the risk he is taking in mentioning such a dangerous topic. For on p. 28D, where the same subject is touched, he says a similar prayer. However, on p. 68A, when starting on the creation of man, although he uses a similar introductory formula, 'let us once more shortly and quickly return to the beginning', the prayer is omitted, which seems to indicate that no esoteric doctrine was touched here.

Evidently, these remarks alluded to some highly esoteric doctrine, probably connected with the idea of an original chaos of 'disordered motion'. There is a fair probability also that this doctrine was a religious one, imported from the East; but the question whether this doctrine was based upon the conception of 'original man', as L. Troje maintained,[1] or perhaps upon that of Mother Earth, the great dark goddess of the mysteries,[2] will probably remain a secret as it was already at the time when Plato wrote the *Timaeus*.[3] The important thing for us is that in the course of these considerations Plato established the abstraction 'matter', *ὕλη*, on a moral, not scientific basis.[4] What happened there has found its classical expression in the very formula which was the apple of contention between Reitzenstein and Schäder, a description of Mani's heresy by the Neo-Platonist Alexander of Lycopolis: *τὴν γὰρ ἐν ἑκάστῳ τῶν ὄντων ἄτακτον κίνησιν ὕλην καλεῖ.*[5] Without any change this sentence may be said to describe Plato's own doctrine of 'matter'.

The further discussions about *ἀρχή* in the *Timaeus*, particularly those concerned with the geometrical 'beginning' of fire as well as with 'the earlier *ἀρχαί*',[6] seem to me to touch only the fringe of the question of original 'beginning'. They appear to have been evoked mainly by the

[1] L. Troje, *Mus. Helvet.*, 1948, 96 f. The article, published during Mrs Troje's last illness, refers to a larger and more fully documented work on the original man, deposited by her at the Göttingen University Library. No steps to ensure its publication seem to have been taken; and the deplorable feud between H. Schäder and R. Reitzenstein on the subject has made it unattractive to other historians of religion.

[2] This idea is suggested by *Tim.*, 50DE, and by Hesiod, *Theog.*, 220. In Orphism Mother Earth was the main subject whom it was *οὐ θέμις* to discuss with the uninitiated, Diodor. Sic. III.62.8, cf. *Tim.*, 51A, where Plato rejects the name of 'Earth' for 'the mother'. A. Dieterich, *Mutter Erde*, 3rd ed., 1926, 55 f., discusses the worship of Mother Earth in the Eleusinian and Orphic mysteries, but unfortunately does not refer to the *Timaeus*. The difference between the 'mother' in *Tim.*, 51A, and the 'nurse', ibid., 49A, 52D, which is abstract matter, should not go unnoticed.

[3] *Tim.*, 53D, *τὰς δ' ἔτι τούτων ἀρχὰς ἄνωθεν θεὸς οἶδε καὶ ἀνδρῶν ὃς ἂν ἐκείνῳ φίλος ᾖν.* Orphic cosmogonies know of a creation by Phanes, preceding that by Zeus, W. K. C. Guthrie, *Orpheus*, 2nd ed., 1952, 80, but I have not succeeded in establishing any definite influence from there upon the *Timaeus*.

[4] *Tim.*, 30A.

[5] Alex. Lycop., ed. Brinkmann, 5, 8. I am well aware that Alexander ascribing this doctrine to Mani, contrasts it with *Tim.*, 49A, in particular; but his understanding of this passage is erroneous (Aristotelian), and the formula is traditional, as Mrs Troje has shown by reference to Plutarch, *De Iside* 64, op. cit., 98.

[6] *Tim.*, 53DE.

rejection of any belief that the *ἀρχαί* of 'disordered motion' should be identified with the four elements. They will mainly serve to emphasize the fact that Plato was rather hesitant in using, as an alternative, the term *ἀρχή* for the description of that rather mystical subject. Professor Skemp has suggested that here, where the influence of the Orphic mystery doctrine was no longer prevalent, traces of the philosophy of Anaxagoras may be found, in which the one *ἀρχή* of the Nous found its counterpart in the innumerable *ἀρχαί* of original molecules.[1] This is, of course, a plausible hypothesis. However, when it is considered that Timaeus, at the beginning of his speech, reduces the material *ἀρχαί* to one ultimate and mysterious 'beginning', something for which no evidence from Anaxagoras seems to be forthcoming, it may be preferable to look for evidence of the Pythagorean doctrine even here.[2]

As the last stage of our enquiry into Plato's conception of *ἀρχή* we shall now discuss the famous establishment of natural theology in the tenth book of his *Laws*. Here once more we are greatly indebted to the researches of Professor Skemp, and we accept his view that 'what appears to be scientific exegesis (in the tenth book of the *Laws*) turns out on closer inspection to be "protreptic" elaboration of what can be found in the *Phaedrus* and the *Philebus* and . . . the *Timaeus*'.[3] We are still in agreement with him, and indeed under very considerable obligation to him, with regard to 'the more remote *ἀρχαί*' of *Timaeus*, 53D, which he has convincingly explained as the mathematical point, generating the one-dimensional line, the two-dimensional geometrical figure, and the three-dimensional stereometrical body.[4] It is, however, impossible to regard, as he wants us to do,[5] either the eternal father and demiurge or the Nous as the *ἀρχὴ κινήσεως*, even if Plato could be proved to have entertained any other conception of the 'beginning' than that of

[1] J. B. Skemp, *The theory of motion*, 1942, 32 f.

[2] Timaeus' argument here aims at the discussion about the number of material cosmoi, be it one, five (cf. the regular 'mathematical' bodies) or an infinity. The subsequent polemics, *Tim.*, 55CD, seem to me directed against earlier philosophers, e.g. Anaximander, rather than against Democritus and the atomists; and this may point to a Pythagorean source.

[3] J. B. Skemp, *The theory of motion*, 1942, 96.

[4] J. B. Skemp, op. cit., 104 f.

[5] J. B. Skemp, op. cit., 109 f. Professor Skemp here claims that the father and demiurge is described by Plato as 'the ultimate *ἀρχὴ κινήσεως*'. Professor Skemp has a much greater as well as more profound knowledge of Plato than I can claim; and I can only state that he has not quoted any evidence for this, and that I, in spite of a strenuous search, have not been able to discover such evidence.

'beginning of motion'.[1] For Plato's firm and often repeated conviction, which appears also in the tenth book of the *Laws*, was that the cosmic soul was the ἀρχή—or rather, as there are two supreme souls: the one ἀρχή—or ordered motion,[2] whilst 'another soul' was the source of 'disordered motion'.[3]

In order to establish the first of the claims which we have just stated, the claim that it is the cosmic soul which Plato, in the tenth book of his *Laws* presented as the 'beginning' of orderly motion, the following passages may be quoted in full:[4]

(*a*) Athenian, In this way: Will there ever be another of these, figuring as the first change? But how? As soon as that is moved by another one, then that other will perhaps be the first mover? Therefore it is impossible. But then if a self-moving mover sets another one in motion, and that one another one, and if thus for a thousand times ten thousand times motion were started, would there be any other 'beginning' of all their motion than that of the self-moving mover? Etc.

(*b*) So we say that the 'beginning' of all their motion, and the first among the un-moved as well as the moved, being that which is moved by itself, is the most ancient and strongest of all motions; but that which is caused by another, and in its turn moves others, comes second.—That is truly said.

(*c*) Cleinias, You say that the conception of self-motion has the identical substance as that name which we all pronounce as 'the soul'?—Athenian, So I say. But, if that is so, can we still complain that we have not been shown that the soul is the first origin and motion of all that exists, and has been, and will be, and also of their opposites, since it has appeared as the cause of change and motion in all things?—Cleinias, No. But it has been satisfactorily shown that the soul is most ancient within the universe, having come into existence as the 'beginning' of motion.

These three passages make it plain that it was the soul which Plato regarded as the 'beginning'. He accepted ἀρχή as a technical term, which had had its place in Greek philosophical terminology from the days of Anaximander, even if so far it had not received a generally accepted definition. He made a special effort to stress the dignity of the term by

[1] At first sight the 'more remote ἀρχαί' in *Tim.*, 53D, might appear in such a light; but J. Stenzel, *Kleine Schriften z. griech. Philosophie*, 2nd ed., 1957, 16, shows clearly that this passage refers to motion = change of quality, and that only its form is due to 'die für den späten Plato charakteristische Mathematisierung der Welt'.

[2] This is rightly emphasized by J. Stenzel, op. cit., 11.

[3] J. Stenzel, op. cit., 20 f., has attempted to explain this Platonic dualism as a purely ethical conception, safeguarding the liberty of moral decision to all souls. I feel certain that he has thus mistaken the effect of Plato's theory for its basis.

[4] Plato, *Legg.* x, 894E–895A, 895B, 896AB.

bringing into prominence in these passages the connotation of rulership, as J. Stenzel has shown.[1] On the other hand, since the empirical cosmos was for Plato a functioning entity, a cosmos of motion, he found it necessary to define and delimit the 'beginning' as the 'beginning of motion', a causal beginning, the beginning of 'life'. The entry of the infinite into the finite world was in the eyes of Plato that 'beginning' which set the cosmos in motion. This recognition demanded two further steps to be taken. On the one hand it was necessary to regard the creative activity as continuous. For the cosmos functions by secondary motion, which will come to a rest unless it is constantly revived; but 'the soul', being a 'self-moving mover', cannot rest, but has to impart to 'the body' of the cosmos that motion which is its existence. On the other hand, the cosmos does not simply go through a system of mechanical motions: it functions. This means that its motions are subject to moral judgments: they are either 'orderly' (*εὔτακτος*) or 'disordered' (*ἄτακτος*) motion. And Plato, being a dualist and regarding this contrast as irreconcilable, proceeded to demand that 'disordered' motion should have its 'beginning' as well in a soul. Continuing our reading in the tenth book of his *Laws*, we find there:[2]

Athenian, So it is then accordingly necessary to admit that the soul is the cause of good things as well as evil, of beautiful and ugly, of just and unjust, and of all the opposites, if we state it as the cause of all things?—Cleinias, How could it be otherwise?—Athenian, If the soul permeates and inhabits all things that are moving anywhere is it then not necessary to say, it even permeates heaven?—Cleinias, How else?—Athenian, One soul, or more than one? More than one, I will reply on their behalf. We cannot put less than two, of whom the one is beneficent, and the other capable of working in the opposite direction.—Cleinias, You have stated this most truly.—Athenian, All right. Thus the soul leads all things in heaven, on earth, and in the sea by its movements that are named, to will, to see, to care, to counsel, to argue truly, falsely, joyfully, painfully, bravely, cowardly, hatingly, lovingly, and all other motions which, akin to them as primary motions, lead the universe, embodying the secondary motions of the bodies, towards growth and decay, separation and fusion, and what follows from those as the warm and the cold, the heavy and the light, the hard and the soft, the white and the black, the sour and the sweet, and the bitter, and to everything by the help of which the soul, with the assistance of the always divine Nous, being god herself, directs all things in rectitude and happiness. Is this correctly stated, or are we still uncertain whether it might not somehow be different?—Cleinias, Not at all.—Athenian, Which kind of soul shall we now say,

[1] J. Stenzel, op. cit., 18.
[2] Plato, *Legg.* x, 896D.

has the control of heaven and earth, and the whole cycle: That which is sensible and full of virtue, or that which has neither? Do you now want me to give an account of that?

Here the argument passes on to the discussion of the divine souls of the sun, moon, and stars, which are shown by their regular movements to be beneficent; and after that to the need for a divine providence, even in the smallest matters. 'The soul that has neither' sense nor virtue, however, that source of disordered motion and of evil, is not discussed at length any more, not even on page 906A, a passage in *Laws* x, which is customarily quoted in this context.[1] One fact, however, appears clearly from the Athenian's argument, that the 'beginning' of the empirical cosmos is a soul, and nothing else, which was placed by Plato as the *ἀρχὴ κινήσεως*, even if this soul may deserve no other characterization than that of a 'disorderly and undefined yet self-moving "beginning", which creates motion', as Plutarch described it.[2] Further indications concerning the character of this 'soul' are omitted and it seems that Plato left the matter intentionally obscure. Professor Skemp is certainly right in saying that in Plato's *Laws* as in his *Timaeus* 'the disorderly motions are a constant undercurrent in the actual cosmos'.[3] It is, however, impossible to say on the evidence available in what way Plato's conception of the 'evil soul' was dependent upon Empedocles' cosmic principle of *νεῖκος*, strife, to which Professor Skemp has rightly drawn our attention.[4] The few traces of the Pre-Socratic development of the conception of *ἀρχή* which we have been able to establish, are nevertheless of great significance for the understanding of Plato's own conception of the origin of the empirical cosmos. For it is evident that Plato

[1] 'For since we are agreed that heaven is filled with many good things, but also with their opposites, although more good than bad, we have to say that the conflict is immortal, and demands a miraculous vigilance. Our allies are the gods together with the demons, for we are the property of gods and demons. We are destroyed by injustice (*ἀδικία*) and pride with thoughtlessness; but we are saved etc.' It cannot be decided whether Plato—similar to Anaximander—regarded 'injustice' as a cosmic event. It is not impossible since 'pride' satisfies the subjective requirement.

[2] Plutarch, *De animae procr.* in *Tim.* VI.2, 1014e, cf. J. B. Skemp, *The theory of motion*, 1942, 59 n.2.

[3] J. B. Skemp, op. cit., 59.

[4] J. B. Skemp, op. cit., 54 n.1, has printed a diagram, designed by K. Ziegler, showing the connections, as Z. saw them, between Orphism, Empedocles, and Plato, separating Empedocles widely from Orphic traditions. I wonder, and would rather suggest the possibility that Attic and Sicilian Orphism had different traditions, and that Plato in *Legg.* x, and in *Tim.*, made a conscious attempt at harmonizing them.

started from the protest against the rationalistic attempt made by Anaxagoras to establish the empirical cosmos as the best world possible, taking its origin from the Nous, which for him, as much as for Socrates and Plato, was the embodiment of all that is good. This protest compelled Plato to take up once more the position which had made the enquiry into the 'beginning' imperative for Anaximander and Empedocles, the position where their own *ἀδικία* and *κακότης* assumed a cosmic importance. From there Parmenides' denial of the reality of the empirical world, where Dike, the goddess of justice, offered the choice between changeless existence and absolute nothingness, a cosmos of thought and a chaos of sham, proved a tremendous temptation. However, it appears to have been the Pythagorean heaven of numbers, which had inspired already the conception of a cosmos of ideas in the mind of young Socrates when he discoursed with 'father Parmenides', which continued to exercise its influence upon Plato's cosmology. For numbers being morally indifferent, they were obviously free from evil.

However, Plato did not feel satisfied with moral indifference, neither was he convinced that Parmenides' equation between the evil and the non-existent, the *μὴ ὄν*, was correct. His dualism contrasted mind and matter, Nous and Ananke, good and evil, as in the *Phaedrus*, so also in the *Timaeus*, and in the tenth book of the *Laws*.[1] What was new, and indeed revolutionary in this his newly found cosmology, was just this contrast of mind and matter. I have stated repeatedly that the philosophical abstraction 'matter' does not appear to have been consciously achieved before the middle of the fifth century B.C., and the contrast of mind and matter seems to have belonged only to the generation of Socrates. Finally it was Plato who identified this contrast with that between good and evil. It cannot be stated too often that the conception of matter cannot be proved to have existed because primitive—and not even so very primitive—thinkers need a material medium, wind, the ether, for a conception of spiritual powers. The chemical definition of gaseous matter can hardly be described as a turning point in philosophy, as some augurs on *ruach* and *pneuma* want us to believe. Neither can we

[1] Cf. J. Kerschensteiner, *Plato und der Orient*, 1945, 68 f., and esp. 77 f. (on *Timaeus*); 98 f. (on *Phaedrus*). The case of the author seems, however, overstated when she maintains, p. 76, 'dass es keine böse Weltseele gibt'. For from the passages quoted above, and esp. *Legg*. x, 896D, 906A, it appears to me that Plato regarded 'the evil soul' as a cosmic power, although a subject of esoteric doctrine, and presumably a *pudendum*. The a uthor's claim that the soul 'ist an sich indifferent' (p. 99 sub 1), cannot even be based upon the *Phaedrus*, but is a modification introduced by the Academy.

omit the remark that the probable, but not I think provable, thesis that even the Pythagoreans proceeded from the assumption of an ethereal substance of spiritual things, can be adduced to show that Theophrastus had a correct appreciation of the Pythagorean dyad as 'assembling the *material* (*ὑλικόν*) multitude around itself'.[1] He rather marks the complete change which was introduced by Plato into Greek philosophy by his fixing of a place for evil: Matter. For in this way the separate function of the 'evil soul', stated by Plato himself, admittedly in a somewhat vague and obscurely restrained manner, was eclipsed; and the real 'beginnings' of the empirical cosmos were transferred to the purely metaphysical realm, where Plutarch established Ananke as 'a third principle and power' between god and matter.[2] The result of such a retreat into pure metaphysics was that the transcendental soul, which was 'self-moving' in an autonomous sense, 'herself god', was degraded, described as an empty vessel either to be filled by the Nous or to be left empty, and only thus enabled to produce good, or evil.[3] It cannot be denied, however, that a rapid progress in this direction had already been started by Plato himself. The human soul, which had still been 'soul' in the full sense of the word in *Phaedrus*, was in the *Timaeus* reduced in rank as the last soul to be created, and suffered a further degradation in the tenth book of the *Laws* by the ascription of 'eternity' to the cosmic soul, and of 'divinity' to the souls of the sun, the moon, and the stars. Yet even these possessed apparently only a shadow of true infinity, seeing that the cosmos itself was essentially finite, continuing its existence only because of the eternal mercy of the father, the demiurge, who had created it.[4]

[1] J. Kerschensteiner, op. cit., 113, referring to H. Diels, *Doxogr.*, 302, 17.

[2] Plutarch, *De animae procr.* in *Tim.* VI.8, 1015c. Ernst Hoffmann, *Platonismus und Mystik*, S.B. Heidelberg 1934/5, 2, 39 n.3, has rightly characterized this definition of Ananke as an attempt at interpreting the Platonic triad of god, soul, and body in the light of later academic philosophy.

[3] Plutarch, ibid. VI.2, 1014e, *αὕτη γὰρ ἦν ψυχὴ καθ' ἑαυτὴν · νοῦ δὲ καὶ λογισμοῦ καὶ ἁρμονίας ἔμφρονος μετέσχεν, ἵνα κοσμοῦ ψυχὴ γένηται.*

[4] I feel that J. B. Skemp, *The theory of motion*, 1942, x f., may have paid too much attention to the claims made by Xenocrates and certain Neo-Platonists, that Plato's aim in producing the creation story in *Timaeus* was purely didactic. Not only Plutarch, and Philo, *De aetern. mundi* 52 f., edd. Cohn–Reiter, VI, 89 f., but also the doxographers, Theophrastus–Aëtius I.7.4; Galenus, *Hist. phil.* 17, cf. ibid. 35, Diels, *Doxogr.*, 299 f.; 609 and 618, both obviously dependent on Aristotle, have treated Plato's report as being sincerely historical. Cf. W. Spoerri, *Späthellenistische Berichte etc.*, 1959, 107 f., who would characterize such reports as the ones here quoted as 'diacrisis cosmogonies'.

And the Academy seems to have secreted—or lost?—all notion of where evil took its 'beginning'.[1]

[1] There is a curious terminological similarity, which deserves to be taken notice of, between Plutarch's report on Plato's doctrine about the origin of evil, *De animae procr.*in *Tim.* VI.8, 1015C, *ἐπεισόδιον οὐκ οἶδα ὅπως ποιοῦντα τὴν τῶν κακῶν φύσιν*, and Paul, Rom. 5:20, 'and the law came in beside' (*παρεισῆλθεν*), and I think that there is good reason to assume that Paul here alluded to an esoteric doctrine, cf. the remarks on *παρέρχεσθαι* in E. Peterson, *Frühkirche, Judentum und Gnosis*, 1959, 297, 301.

7 Aristotle

Before attempting to outline Aristotle's conception of the 'beginning', which in many respects became final, it may be worth our while to sum up the position reached by Plato. Here we have to begin with two negative statements which for him are fundamental. The first is that, however deeply he may have been under obligation to the Pythagoreans, he never suggests in his dialogues that he adopted their thesis that it was number which should be described as the 'beginning'. This in its turn is confirmed by the fact that for Plato the 'beginning' is not static, but beginning of motion; and that beyond this there are mysterious powers which are not to be included in the scheme of his world. The 'beginning' in Plato's philosophy is thus not a principle, but an event. This view brought about his second negative conclusion, that the row of 'beginning, middle, and ending' is incapable of being used for the description of the empirical cosmos as such, however useful it may be for the description of processes within this cosmos. For whilst it may be assumed that within the limits set by the cosmos the sentence, wherefrom his beginning is, thereunto his end shall be also, has a meaning, the general rule is rather that those affairs will come to a finish in which the beginning and the ending are not combined. For gyratory motion, the motion as such, is the real 'beginning', and this 'self-moving mover' is the soul.

This is where something new was introduced by Plato, and introduced in a new way, into Greek philosophy—the way of myths. Quite apart from the substance of these myths, which, at any rate in so far as the soul is concerned, is astral in character already in the *Phaedrus*, as has been shown by J. Stenzel,[1] it is the mythical form, or method, which is of importance. If it is true, as we have tried to show, that the way of Greek philosophy, from Anaximander to Parmenides without any doubt, and also to Anaxagoras and Democritus, was that of the rational separation of philosophy and mythology, Plato found himself compelled—inevitably compelled—to accept the challenge of mythology, and to establish its interaction with philosophy. For his predecessors had met with demonstrable failure. The rationalism of Anaxagoras had led to the unacceptable conclusion that the empirical cosmos was 'good', and even

[1] J. Stenzel, *Kleine Schriften z. griech. Philosophie*, 2nd ed., 1957, 1 f.

the best possible world. Here, however, not even the question could be answered how from his scientific premisses the introduction of such ethical terms as 'good' and 'best' could be justified. The neglect of Democritus by Plato resulted in the end from the fact that all philosophical materialism is faced with the paradox that it has to seek for a law of chance, whereas human logic demands that chance should be excluded by law. Finally, the absolute determinism of Parmenides led to the paradox that its basis, the claim that 'being and thinking are identical', could not say anything about the object of such thought. For if this object was 'being' then the very subject was its own object, and the result a meaningless tautology, either 'thinking is thinking about thinking' or 'being is being'. If, however, the object of this thinking was the non-being, the followers of Parmenides were led on to break the command, which in the pursuance of his system Parmenides had given, prohibiting all consideration of the *μὴ ὄν*. By the use of the mythical form Plato protected his philosophy from the pitfalls of paradox; but he deprived it of conclusiveness. He achieved a system of thought; but even at its best—and the myth in *Phaedrus* can hardly claim this distinction—it was a very open system.

It may have been noticed that in this short outline of Plato's cosmology, by drawing a demarcation-line against his predecessors, one of their philosophical systems did not qualify for an epigrammatic criticism: Pythagoreanism. We have stated that Plato did not adopt their claim that number was the 'beginning', just as he, rejecting the much too mystical formula *ἓν τὸ πᾶν*, denied this designation to the One. Nevertheless, theirs was not a single-minded approach. Even within the restricted scope of our research their traditional combination of mathematical rationalism with Orphic symbolical mysticism has come clearly to the fore. Plato's connection with Pythagorean mathematics is an established fact. But it is still, largely for lack of sources, a matter for debate whether or not his myths were drawn from the same source. We have referred to the Orphic myth of a creation by Phanes as a possible analogy to Plato's 'evil soul' creating a disorderly motion, already at the very time when the 'father and demiurge' obtained permission to reform by the 'good soul' this primeval chaos.[1] Beyond such an immediate source, Iranian influences upon Plato seem highly probable, whatever the final outcome of the conflict between Reitzenstein and Schäder may be.

It was at this impasse that Aristotle came in; and we feel that the certain animosity against the Pythagoreans, which has appeared in his

[1] Cf. *supra*, p. 99, n.3.

reports about them, may be taken as a pointer to the issue between him and his teacher. F. Solmsen[1] has quite strongly and correctly pointed out that, at least for Aristotle,[2] 'the reality or non-reality of genesis is not by itself a cosmological problem', but that 'it is an issue of a more general and theoretic type'. However, dealing extensively with this 'physical' problem, as he calls it, he has nowhere analysed the meaning given to *ἀρχή* by Aristotle, and has thus somewhat impaired the valuable, and at times provocative, statements both on Plato's and Aristotle's 'principles' of their respective cosmologies.[3] For it is true to say that throughout his extant work Aristotle seems to have battled with the conception of the 'beginning', and this fact can now be easily established by reference to the careful collection of sources in the examination of chance and causality in Aristotle by Helene Weiss.[4] There can be no doubt that the impression created by Solmsen of the novelty of the contrast between 'spirit' and 'matter' is a true one; and that in particular the conception of matter as the substratum to reality is a revolutionary step away from earlier 'natural' philosophy. On the other hand, an inspection of the use made of *ἀρχή* by Aristotle may assist us to appreciate both the help and the hindrance provided by the Pre-Socratics whom he quoted so abundantly.

In this respect Miss Weiss has, unfortunately, limited her remarks to the somewhat vacuous statement that Aristotle by combining the concept of *ἀρχή* with those of *στοιχεῖον* and *αἰτία* has clarified and enriched the problem as seen by his predecessors. Even if this verdict about Aristotle's thought on the subject of the 'beginning' would provide an adequate outline of it, such limited praise would not do justice to the result achieved. For in a certain sense the fact that an ultimate cause, or set of causes, has to be placed at the 'beginning' in a cosmos which is understood as functioning, even living, had been agreed upon already by Plato and Anaxagoras. And again in a certain sense it can be held that the 'elements' as understood by Empedocles, may be seen in analogy to the transition from the 'substratum', matter, to the existing reality of

[1] F. Solmsen, *Aristotle's sytem of the physical world, Cornell Studies in class. Philol.* xxxiii, 1960, 6.

[2] Solmsen has left the question unanswered whether a theory of a *creatio ex nihilo,* as that by Xeniades, has to be treated as the basic issue of cosmology rather than as a 'physical' one, as he calls it. For such a theory puts the idea of 'physis' into doubt by denying the *ποιητικόν* a corresponding *παθητικόν*. However, this question need not worry us here.

[3] F. Solmsen, op. cit., 40 f., 74 f.

[4] H. Weiss, *Kausalität und Zufall in der Philosophie des Aristoteles,* 1952, 11 f.; 14 f.

the cosmos, which Aristotle on occasion found in the four elements, pronouncing them as the passive 'beginning'.[1] Incidentally, it may be considered doubtful whether his exposition added very much to the clarification of the concept of the 'beginning' amongst the Pre-Socratics or rather obscured it. To my mind it has been the main cause of that misunderstanding of Pre-Socratic philosophy which has prevailed from the days of the doxographers unto our own time—and not only, as Miss Weiss would have it,[2] during the nineteenth century.

Aristotle rightly regarded the question of the 'beginning' as belonging to his first philosophy. Thus, although the discussion of the meaning of ἀρχή in his *Physics*, so ably analysed by Miss Weiss, is of great importance for us, and will assist us in the course of our investigation, the problem as such is a metaphysical one. It also figures largely in his *Metaphysics*, and a warning has to be issued at once that the final solution of this problem must not be sought for in the first three chapters of Book Four only, although the 'beginning' forms there the main subject of the discussion. We would rather maintain that the first five books of the *Metaphysics* are shot through with repeated considerations of the 'beginning', and that it is these considerations, which have time and again been regarded as separate from, and even alien to, Aristotle's main purpose,[3] while in fact they form the real thread running through his arguments in these books.

[1] Cf. esp. F. Solmsen, op. cit., 77 f., 121 f.

[2] H. Weiss, op. cit., 13, 'whether the early Greek philosophers ever approached the question of ἀρχή only in this sense (viz. that 'beginning' = element) is something that cannot be proved. Nineteenth-century exegesis may have been too prone to find none but this meaning in their sayings.' It seems to me that this statement begs quite a number of questions: on the one hand, whilst Aristotle ascribed to one or other of the earlier philosophers the view that one of the four elements was to be regarded as ἀρχή, it is yet certain that the contrast between the four elements and the κυρίως ἀρχαί is post-Aristotelian, cf. Empedocles frg. A 28, Diels-Kranz, 5th ed., I, 287 f. (Theophrastus); on the other hand, it has to be said that the time when the doctrine of the four elements was firmly established in Greek natural philosophy is quite uncertain, but rather the fourth century than the fifth. As regards the nineteenth century, I find the meaning 'element' for ἀρχή mentioned in the dictionaries by Passow (5th ed., 1841), and Pape (3rd ed., 1880), but without any reference. A monopoly for this meaning is not claimed even in philosophical speech, and the meaning of 'principle' is given the preference in both these dictionaries.

[3] Reference has to be made here to the instructive review of the various theories about the evolution and compilation of Aristotle's *Metaphysics*, which may be found in M. Wundt, *Untersuchungen zur Metaphysik des Aristoteles*, 1953, 11 f., although it must be said that the author's views are somewhat biased.

However that may be, and at the risk of being proved wrong by a more thorough examination of the coherence of these books, it has to be stated that it would be a mistake to assume that even here only one precise conception of ἀρχή prevails; and the mistake would be even greater if the attempt were made to get at such a conception in the hope of establishing the view to which Aristotle adhered in his early period.[1] Whether or not it is true to say that parts of these books go back as far as the time of his stay at Assos, 347–344 B.C., it seems evident that already in these books one fundamental change in his conception of the 'beginning' makes its appearance repeatedly, the doctrine of an 'unmoved mover as the beginning of existence'. H. v. Arnim has established this fact, which he uses, rightly I think, as test for the establishment of the relative age of the various statements in *Metaphysics*, treating as earlier those which show no trace of this doctrine.[2] For since this doctrine is clearly directed against basic views held by Plato, especially in his last years, and since it is well known how long it took Aristotle to establish his independence against his master's philosophy, it has to be ascribed to the time of his teaching in the Lyceum (after 334 B.C.), whereas the other statements must belong to an earlier period in so far as they are logically incompatible with it. But even those which simply neglect it are likely to fall, under the same verdict. It is, therefore, necessary to subject all Aristotle's statements on the 'beginning', which occur in his *Metaphysics*, to the test whether or not they conform to this doctrine; and it is not advisable either to omit this test with regard to his other writings. For it might lead to serious misrepresentations of his views if a statement which shows clear traces of the theory were used for the interpretation of one which shows none, and is therefore, presumably, of a considerably earlier origin. Viewed from this angle, the approach made by Miss Weiss, which shows no signs of these necessary precautions, may be regarded as somewhat too harmonistic.

This, however, is by no means the only difficulty which will be encountered by those who attempt an investigation into Aristotle's conception of ἀρχή. In addition to his established change of mind, there is always the possibility that alien additions may have crept in before Andronicus of Rhodes finally established our corpus of Aristotelian

[1] W. Jaeger's view, approved by K. Prächter in Überweg–Prächter, *Die Philosophie des Altert.*, 12th ed., 1926, 368, that *Metaph.* I–V contain in the main the earlier parts, is perhaps not altogether safe to hold.

[2] H. v. Arnim, *Die Entstehung der Gotteslehre des Aristoteles*, Acad. Vienna, 1931, *passim.*

writings.[1] There is finally the fact that Aristotle's dialectical method tended to cause him to make conditional statements. Such statements may or may not contain his own convictions. As an example a passage from the *Metaphysics* may be quoted, which very closely touches on our problem of the 'beginning'. There we read:[2] 'Thus it is common to all "beginnings" to be the first out of which something exists or comes into being or is recognized.' This short sentence has to be appreciated critically in two respects before it can be used in the description of Aristotle's doctrine of the 'beginning': firstly, it is to be noticed that the time factor still plays a part in this definition of *ἀρχή*. This is of historical importance. For a hundred years ago, F. A. Trendelenburg already established the fact that at a later period Aristotle expunged it from that definition.[3] Secondly, the sentence leaves us entirely in the dark as to whether Aristotle distinguished between an ultimate 'beginning' and secondary 'beginnings', and this uncertainty is increased when the context is considered. For it is only a little further on that he continues:[4] 'Therefore is nature a "beginning", and element, and intention, and motive, substance, and reason why.' This is a somewhat heterogeneous collection, brought together mainly, so it seems, to enlarge upon the introductory claim that 'all causes are *ἀρχαί*';[5] but as such it may contain implicitly the opinion that indeed there is no distinction between one 'beginning' and another.

Such statements may be seriously misleading when they are logically pressed, e.g. when it is asked whether or not they are reversible. For it is only much later in the same book of the *Metaphysics* that we find a type of 'beginning' which cannot easily be considered as a cause. This

[1] It is here that we have to join issue with the attitude taken by P. Gohlke, *Die Entstehung der aristotelischen Prinzipienlehre*, 1954, 3 f. For if it is correct to treat Aristotle's *Physics* as a genuine production, coming from the Stagirite's pen—and I have to state that J. Zürcher, *Aristoteles Werk und Geist*, 1952, has not succeeded in undermining my confidence—it is unreasonable to ascribe to him two conflicting passages in close vicinity, as *Phys.* VII, 274a, 1–9, and ibid., 28 f. Even if it is admitted that Aristotle's esoteric writings showed emendations and corrections in the master's own hand, it is hard to believe that they contained such blatant inconsistencies. If, therefore, Gohlke's proof is accepted that the first passage is Aristotelian, then the second must be pronounced spurious, as was done already by Immanuel Bekker.

[2] Aristotle, *Metaph.* IV.1, 1013a, 17 f.

[3] F. A. Trendelenburg, *Aristotelis de anima*, 2nd ed., 1877, 157, commenting on *De anima* I.1, 402a, 6.

[4] Aristotle, *Metaph.* IV.1, 1013a, 20 f.

[5] Ibid., 1013a, 17.

is the case when it is stated that 'ability' (*δύναμις*) is to be described as the 'beginning' of motion or change, either when existing in another person or as 'something alien'.[1] For ability in the cases of the architect and physician, adduced by Aristotle, may be described as *conditio sine qua non* for the result, the house or the health of the patient, but hardly as its cause in a specific sense. And the same may be argued when it is also said that the capability to endure (*τὸ δυνατόν*) 'may in one respect be described as holding the "beginning" of motion or change (for even the static possesses capability) in another subject or in itself quâ another subject'.[2] These two remarks make it clear that the 'beginning' was a wider conception than cause—in this context. For the difficulty remains whether or not these latter statements are contemporary with the former. Tentatively it might be considered that the certain emphasis shown in the equation of 'motion or change', which may signify that motion is subsumed under change—and not *vice versa*, as by Plato—is a hint at the theory of the 'unmoved mover', and therefore these two quotations from *Metaphysics* IV. 12 are of a more recent origin than the preceding one from *Metaphysics* IV. I. If this hypothesis should be true it might move one more prop from under that imposing construction of *νοῦς ποιητικός* and *παθητικός* which has played such an important part in the history of the interpretation of Aristotle's philosophy.[3]

Finally, there appears in Aristotle's remarks on the 'beginning' a certain reserve and ambiguity comparable—if much less strongly marked—to that which we have encountered in Plato's references to the 'evil soul' as a second ultimate 'beginning'. These are transcendent, mystical matters which even Aristotle treated with awe. Such a reserve seems to play its part e.g. in the following statement:[4]

> Thus it appears that as often as the 'beginning' is mentioned the boundary (*πέρας*) is mentioned also, and even more often. For the 'beginning' sets a boundary; but the boundary in its entirety is not the 'beginning'.

Once more, what may be gained from this saying for Aristotle's understanding of the 'beginning' comes to us by implication rather than by plain statement. If we imply, justly I think, that Aristotle's point of departure was the belief that one defined finite phenomena by the establishment of their 'beginning', this sentence may be wholly referred to

[1] Aristotle, *Metaph.* IV.12, 1019a, 15 f.

[2] Ibid., 1019a, 22 f.

[3] Cf. the valuable thesis by H. Kurfess, *Zur Geschichte der Erklärung der aristotelischen Lehre vom sog. Νοῦς ποιητικός*, Tübingen, 1911.

[4] Aristotle, *Metaph.* IV.17, 1022a, 4 f.

secondary 'beginnings'. However, the last clause seems to betray that the statement should apply also to the cosmos as a whole; and that it is in fact akin to statements like that of Melissus, that all that exists, exists from the 'beginning'. It seems legitimate to draw from this sentence the conclusion that Aristotle regarded the empirical cosmos in its totality as a finite system,[1] with the 'beginning' taking its place on the boundary, i.e. being transcending without setting out the whole circumference of the limits of the cosmos.[2] He did not indicate, however, what exact place on that boundary of the cosmos was occupied by the 'beginning', and in what way such a special position for the 'beginning' would be marked out.

For this introduction into the questions raised by Aristotle's entering upon the problem of the 'beginning' we have concentrated upon passages taken from the fourth book of his *Metaphysics*. The reason for this is twofold. On the one hand it may be said that nowhere else in this or any other of his works do we find the word ἀρχή as frequently used as here. *Metaphysics* IV can in this sense be claimed as the *sedes materiae* for our investigation. On the other hand, Aristotle appears here in a special way open to the difficulties presented by the notion of the 'beginning'. The passages quoted all reflect upon a conception of ultimate 'beginning', but make hardly any attempt at determining its meaning. Rather is it true to say that they are simply based upon the analysis of the numerous popular uses of ἀρχή collected in the first chapter of *Metaphysics* IV. I hesitate to approve in general of the statement, 'it is Aristotle's characteristic way to take his terms from popular living speech, and by defining them more precisely to elucidate that which the pre-philosophical speech already intended as their meaning';[3] but it is borne out for the word ἀρχή within the context of *Metaphysics* IV. However, it must not be forgotten that side by side with this method there went Aristotle's most careful use of the work of his predecessors. The fact has to be stressed that without Aristotle's efforts the fairly exact and

[1] Starting from quite different propositions, F. Solmsen, *Aristotle's system of the physical world*, Cornell Studies XXXIII, 1960, 160, has quite rightly stated, 'the infinitely large would, if it were real, seriously interfere with Aristotle's cosmological doctrines'.

[2] This view seems to me to conflict with that held by Anaximander, rather than to be derived from it; and I find the origin of this deviation in Aristotle's introduction of the concept of 'matter as such'. So much for the not fully guarded, but important, remark by F. Solmsen, op. cit., 118 f., esp. 121.

[3] This view is fairly commonly held. The quotation is from H. Weiss, *Kausalität und Zufall*, 1952, 12.

comprehensive information about the use of the word ἀρχή and the concept of the 'beginning' in Pre-Socratic thought would not have been at our disposal. It is to be assumed, therefore, that *Metaphysics* IV signifies Aristotle's rejection of all these earlier attempts at giving a precise philosophical denotation to the word ἀρχή. It appears to us that at some point of his career, and presumably in its later (latest?) stages, the aged Aristotle felt that a new beginning with the 'beginning' was required of him. We regard it as a sign of deep resignation when we see the man to whom we are indebted for almost all the information about the earlier philosophical use of ἀρχή, basing his own argument upon the various uncritical usages in which the word makes its appearance. We feel that in his earlier days he would, and in fact did, boldly place the word in the place where it belonged, in the centre of his ontology. We also believe that the point at which he abandoned this attempt, and resigned himself to a mere description of the various meanings of the 'beginning', can be clearly shown in the development of his philosophy.

The tenth book of Aristotle's *Metaphysics* appears to me the most enigmatic part of an enigmatic work. Even within the loose texture of the whole work, this book seems to interrupt the sequence wherever it is given a place. In former time, therefore, various attempts were made to prove the spuriousness of at least its first half, chapters one to eight; and I consider it the great merit of Werner Jaeger to have shown the correct approach to the book. He has made it very probable that it should be treated as the valuable testimony to an earlier, still Platonizing stage in Aristotle's philosophical development.[1] Even the most determined—and not altogether fair—attack upon Jaeger's views on the growth of Aristotle's philosophy has conceded this point,[2] which for the time being may be treated as *communis opinio*. The isolated protest by P. Gohlke,[3] who suggests that the tenth book of the *Metaphysics* is its most recent part, is as yet no more than a rather biased, though able, reminder that the riddle of the tenth book has still to receive a complete solution; but as it stands it is unacceptable. From the same camp, however, comes a much more substantial contribution, which it seems easy to align with W. Jaeger's statement. M. Wundt[4] has offered proof that the fifteen—or twenty-seven[5]—*aporiae* proposed in the second book

[1] W. Jaeger, *Aristotle*, transl. R. Robinson, 1948, 208 f., with a discussion of earlier views, and in particular of Natorp's claim that chapters 1–8 are spurious.

[2] H. v. Arnim, *Wiener Studien* XLVI, 1929, 13 f.

[3] P. Gohlke, *Die Entstehung der aristotelischen Prinzipienlehre*, 1954, 12 f.

[4] M. Wundt, *Untersuchungen etc.*, 1953, 33 f.

[5] M. Wundt, op. cit., 30.

of the *Metaphysics*, issue in the last resort from one central one, that about the ἀρχαί. Wundt's proposition seems sound enough, and is therefore of great importance for us. For it shows that at a time somewhat earlier than that when *Metaphysics* IV was conceived, the question of the 'beginning' figured as the basic *aporia* in Aristotle's metaphysics. We may thus conclude that before he took the step of dealing with the 'beginning' in a pragmatist way in *Metaphysics* IV, he had found it his supreme *aporia* in *Metaphysics* II.

This book in its turn, as we have seen, runs in various respects parallel to Book Ten, but in such a way that it shows its dependence upon that book. This is true in particular with regard to the question of the 'beginning'. For whilst this question is spread over the whole of Book Two,[1] it is discussed in a concentrated form in the second chapter of Book Ten. Since, on general grounds Book Ten has to be regarded as the earlier of the two, we may conclude that Aristotle had this chapter clearly in mind when writing *Metaphysics* II, and that it also represents the earliest stage of his considerations about the 'beginning', which may be reached by the literary criticism of his *Metaphysics*. For this reason *Metaphysics* X.2, seems to be the natural starting point for the examination of his doctrine:[2]

§ 1. If, however, the principle sought for now is not detached from the bodies, what else could one propose, but matter? Matter, however, does not exist actually, but potentially; and thus the idea and the form would appear as a higher principle. Yet these are perishable, and hence there would be altogether no eternal substance existing separately and by itself. That, however, is unacceptable. For it has been recognized as a subject of independent enquiry by all the finest intellects that such a substance and principle exist. For how could there be order if there were nothing eternal, separate and everlasting?

§ 2. Moreover, if there is such a substance and a principle of such a nature as we now try to establish, as the one identical principle of the things eternal and perishable, we are faced with the difficulty why on the basis of the same principle the

---

[1] Aristotle, *Metaph.* II.1, 995b, 5 f., describes as the ἀπορία πρώτη the question whether it is the only task of philosophy to enquire into first principles (ἀρχὰς πρώτας), or into all 'those which form the basis for any kind of proof'; and from there onward it is always ἀρχαί, principles in the plural, which form the subject of discussion in the subsequent chapters. This is a significant deviation from *Metaph.* X.2, which starts with examining ἀρχή in the singular, a conception which is absent from *Metaph.* II altogether. It must therefore not be assumed that Book II contains no more than an application pure and simple of such results as had been obtained in the course of Book X.

[2] Aristotle, *Metaph.* X.2, 1060a, 13 f. to 1060b, 30 f.

one part of things subject to this principle is eternal, but not the other: for that is unacceptable. If, however, there is one principle for things perishable and another for things eternal, if the former is eternal the same difficulty arises: Why, if the principle is eternal, are not the things subject to it also eternal? If, however, it were perishable there would be another principle of that, and yet another and, so on *ad infinitum*.

§ 3. If again one posits those two which most of all appear to be immovable principles, the One and Being, as principles, the question arises first of all how they, since neither of them signifies a definite thing (τόδε τι) or a substance, will be separate and existing by themselves? However, the eternal and first principles sought for must be so qualified. Yet if the two should be definite things and substance, all things are substance; for of them all it is claimed that they exist, and of some that they are one. Nevertheless, it is a false claim that all things existing are substance.

§ 4. Again, how can it be accepted as true when it is claimed by some that the One is the first principle, and also substance, and who out of the One and of matter produce the first number and say that this is substance? For in what way can the dyad and all the other composite numbers be thought of as a One? Neither do they say anything about this, nor would it be easy to do so.

§ 5. If then one would claim that the lines and what they enclose (i.e. the regular geometrical figures) are first principles, they are not detached substances, but sections and divisions, the lines of figures, the figures of bodies, [the points of lines,] and also the limits of the said things. They are thus contained the one in the other, and none of them is separable. And how can one regard the One and the point as a separate substance? For of all substances there is a coming into being, but not of the point. For the point is a dissolution.

§ 6. There is also the difficulty that every science deals at one and the same time with fundamentals as well as with a particular subject; yet the substance is not a fundamental, but rather a substance of something. Therefore, if there is a science of first principles, in what sense should it be admitted that the 'beginning' is a substance? Further, whether there exists anything beyond the sum total, or not? If not, the universe, being material, is perishable; but if it is so, one has to think of idea and form (as being transcendent). Of these, however, it is hard to say for which objects they exist, and for which other ones not. In some cases at any rate it is evident that the idea has no separate existence, e.g. in the case of a house.

§ 7. Finally, there is the question whether the first principles are identical according to their idea or to number; for if they are so according to number then all things would be identical.

It has been necessary to quote this passage in full in order to show the determination with which Aristotle set out to master the conception of ἀρχή, proposed so daringly by the earlier philosophers. At the same time it should also appear that the passage quoted may not be altogether homogeneous, but reflect a development by stages. Thus it would not exhaust the meaning of each paragraph if it were simply ascribed to some philosophical tradition, as e.g. § 3 to that after Parmenides, or § 4 to that after the Pythagoreans, even if it could be shown in that way that all its arguments are drawn from earlier philosophical discussion. Only one fact has to be stressed, that it is evident that here Aristotle did not base his discourse upon the common usage of the word *ἀρχή*, as in the first chapter of *Metaphysics* IV. In X.2 the philosopher's criticism is rather directed towards the traditional philosophical conception of *ἀρχή*, and there is no attempt made to question the terminological aptitude of the word for defining the ultimate origin of the world, as in IV.1. We may conclude then that at the time when X.2 was written such considerations had not yet come to Aristotle's mind.

This conclusion is strengthened when the relation between our passage and *Metaphysics* II, the book of *aporiae*, is shortly scrutinized. It is characteristic for Book II that in it the promise is made to analyse one or other of the *aporiae*, and that such promises are carried out, at least in a number of cases, in the later books of the *Metaphysics*. It cannot, however, be maintained that *Metaphysics* X.2 should figure amongst these cases. Its ethos is different. For the intention in this chapter is obviously not that of solving difficulties, but to uncover final antinomies of the human reason.[1] Logically *Metaphysics* X.2 has its place before, not after, the book of *aporiae*. It is true that one or the other antinomy stated in our passage is discussed also in one of the earlier books of the *Metaphysics*;[2] but in such a case an elaborate dialectical approach, which is

[1] M. Wundt, *Untersuchungen etc.*, 1953, 35, has drawn a short comparison between Aristotle's and Kant's dealings with these antinomies, without however making any distinction between Aristotle's method in *Metaph.* X, on the one hand, and II, on the other. This is to be regretted, and Wundt's statement, 'Aristotle is concerned with the two methods of thought and doctrine the influence of which had been decisive for his own development, the Ionian to which he had an innate leaning, and the Platonic in which he had been instructed', smacks of racialism, and will not meet the case of *Metaph.* X.2. However, he has made a worthwhile contribution by his discussion of Aristotle's real and apparent solutions of the *aporiae* in Book II, which follows immediately after this sweeping statement.

[2] P. Gohlke, *Die Entstehung etc.*, 1954, 14, remarks on the statement concerning the point made by Aristotle in § 5: 'Nobody can understand this without having read the more explicit explanation in II, 1002a, 30 f.', and follows up this warning with his own—incorrect—interpretation of § 5. He also proclaims that there exist no parallels between *Metaph.* IV and X, but does not offer any reason why *Metaph.* IV.1, dealing with the same subject, keeps silent about X.2 or *vice versa*.—The reason why our § 5 seems perfectly intelligible in itself, lies in the fact that its attack upon the

absent from our passage, seems to witness to the previous existence of the proposition so discussed. This may be regarded as yet another sure sign that such analogous texts are dependent on the already stated antinomies of *Metaphysics* x.2.

All these observations joined together show, therefore, that this chapter contains indeed the earliest attempt made by Aristotle at grappling with the problem of the 'beginning', to be found in his *Metaphysics.* The question remains why this attempt was formulated as a collection of unresolved questions, although earlier philosophers, even Plato, had been confident that they had found the answer? M. Heidegger is said to have maintained in his lectures on Greek philosophy, given at the university of Freiburg i. Br. in the twenties of this century, that its central problem, down to the time of Aristotle, had been the ontological problem.[1] Whether or not such a radical formula should be accepted in its totality may well be open to doubt. However, it has its use because it shows the outstanding importance of our passage for the understanding of Aristotle's philosophy. At the time when *Metaphysics* x.2 was written he wrestled with the problem whether his particular question, What are we? depended upon an answer to the other, Where do we come from? or *vice versa.* We also learn from our passage that from the premisses of his philosophy he despaired of finding an answer to the question, where do we come from. In this respect § 6 is of particular importance, since it states that an inductive, 'scientific' approach to the problem of an ultimate 'beginning' is logically impossible. Here we find Aristotle laying the foundation for the pragmatist approach to the problem. His reasoning is that the inductive method cannot discard the observation from which it has started without discarding its results as well, and therefore cannot establish the fundamentals upon which the earlier philosophers had so confidently erected their theories. This he follows up in § 7, the last paragraph, by stating that the deductive, mathematical method in its turn cannot proceed to specialized observations, since the identical character of 'number' in all numbers conflicts with the diversity of the empirical world.

Eleatics, which originally had been Plato's, depends of course upon Aristotle's doctrine of continuity; and if the influence of *Metaph.* x.2 § 1, is absent in *Metaph.* IV, it is at any rate evident in I.3, 984a, 17, and II.1, 996a, 9 f., where the categorical demand of § 1, that the ἀρχή must exist actually, and not only potentially, is put into question. Of less importance is the analogy between § 2 and *Metaph.* I (elen.) 3, 994a, 1 f., since it is presumably post-Aristotelian.

[1] H. Weiss, *Kausalität und Zufall,* 1952, 52 n.6, accepting the claim made.

These two paragraphs, therefore, show the beginning of Aristotle's pragmatism; but they may at the same time serve as a background to Plato's myth in the *Timaeus*. They not only make it clear how Plato had been compelled to devise his cosmogony in such a poetical form in order to escape from the barrenness of Sophistic rationalism; but they also give an insight into the fact that Plato's predecessors, in so far as they treated the problem of the 'beginning' with due sincerity, were aware of the inescapable fact that its origin lay in mystery, perhaps in the mysteries, as has been pointed out before. In themselves, however, these two paragraphs form a fitting ending to a passage which is still closely akin to that scepticism which produced an allegorical interpretation of the *Timaeus* already in the early Academy, as we know in particular from Xenocrates.[1] In so far as our passage is concerned, Aristotle appears as being still very far from the fervour with which he came to attack the *Timaeus* in some of his other writings.[2]

Having arrived thus at a relatively certain date for *Metaphysics* X.2, we may now consider its purpose. It seems that already the way of the early Academy led to an increasingly radical scepticism, and in consideration of its date our chapter may be regarded as a witness to this tendency. Whilst I believe this to be a relevant observation, it is nevertheless insufficient for a definition of the character of our chapter. Neither will this definition be complete when it is added that § 4 contains an ostensible attack upon Speusippus, and that there is a considerable possibility that §§ 5 and 6 may be referring to contemporary debates which are known to have taken place in the early Academy. These facts may confirm the date of our chapter as near the time when Aristotle left the Academy, but beyond that they are of little assistance. For neither of these statements accounts for the diversities apparent within the chapter or offers a clue to the philosophical aim it pursues.

The two chief diversities appearing in our chapter are first the change from the singular *ἀρχή*, so strongly emphasized in its first paragraph, to the plural *ἀρχαί* in the subsequent paragraphs. The significance of this change will be discussed later and in a larger context.[3] The second

[1] Cf. Xenocrates frg. 54, Heinze. Überweg-Prächter, *Philos. d. Altert.*, 12th ed., 1926, 345, stress the fact that a similar attitude was also taken by Speusippus and Crantor, whilst Aristotle took the Timaeus myth more literally. Since it will be seen that *Metaph.* X.2 contains a decided attack upon Speusippus, this statement may not only provide us with a date for the passage, after the separation from Xenocrates 348/7 B.C., but may even make us wonder whose interpretation followed Plato's intentions more faithfully.

[2] Cf. in particular *De generatione* II.1, 329a, 5 f.

[3] Cf. *infra*, p. 123.

is to be found in the fact that only some of the problems stated here are taken up again in *Metaphysics* II, the book of *aporiae*, whereas the others have been passed over. This selection indicates that there was a common purpose underlying our chapter, which Aristotle later abandoned. This common purpose is, I believe, easily discoverable. It appears from § 1 that Aristotle had no intention of chiding those 'finest intellects' whom he quotes there, for their attempts at discovering an ultimate 'beginning' of the empirical world. Admittedly, their endeavours had failed so far, but it was right and indeed necessary that they should continue. Therefore a summary of earlier attempts, and the reasons for their failure, would help to clear the way for new ones to be made. If it is true that our chapter marks Aristotle's defection from the Academy, and that it contains polemics against its members, it still bears a sign of loyalty towards its master himself, though a tacit one.

For it is remarkable that the chapter contains no reference to Plato's doctrine that the *ψυχή* is the ultimate *ἀρχή*. Nevertheless, even at this stage, the distance from this view to Aristotle's uncertainty about the ultimate 'beginning' was considerable. That may be illustrated by the first paragraph of his *De anima*, where we find the somewhat hesitant statement that 'the soul is, so to speak, the *ἀρχή* of living beings'. At this time at any rate, Aristotle did not give any great prominence to the idea of the 'beginning' in his doctrine of the soul.[1] It is true that the spheres of the sky are assigned each its own peculiar mover in the late part of the *Metaphysics*,[2] and these in their turn are put in analogy to the soul in the *Physics*.[3] However, the human soul which for Plato, at least in his *Phaedrus*, had been the starting point does not, in Aristotle's view, move in these august circles. According to *De anima* it is only 'so to speak' a principle, and not even regarded as immortal.[4] The fact that these considerations on the soul are complementary to the other remarks in *Metaphysics* X.2 is evident. Aristotle would not—could not?—bring them into an equally epigrammatic form. With regard to a less central theme of the *Timaeus* he has delivered his attack fairly plainly in

[1] Aristotle, *De anima* I.1, 402a, 5. W. Jaeger, *Aristotle*, transl. R. Robinson 1938, 39 f., has shown how closely Plato's doctrine of the soul was followed in Aristotle's *Eudemos*, and how greatly he modified it in *De anima*. It might seem tempting to connect this change with the theory, evolved by the late Aristotle, of the 'unmoved mover' by which he sought to replace Plato's conception of the 'self-moving mover'. However, such an hypothesis would not be very safe.

[2] Aristotle, *Metaph.* XII.8, 1073, 32 f.

[3] Aristotle, *Phys.* IX.6, 259b, 28 f., cf. J. Moreau, *L'âme du monde*, 1939, 115 n.4.

[4] W. Jaeger, op. cit., 45.

§ 1 of our chapter, explaining that the 'beginning' is not to be found in matter as such. The essence of his criticism can be gathered from here, but its full impact and direction becomes clear once more from a more specialized—and I think later—treatise, that 'On coming into being'.[1] Here he takes Plato to task first, because he did not state in detail how exactly matter as such could be thought of as being corporeal, and therefore divisible, which is to be demanded in the 'beginning' of corporeal things; and secondly, because he gave no indication at all about the possible link between matter as such and abstract geometry, a criticism akin to that directed against Speusippus in § 4 of *Metaphysics* x.2. And he himself suggests as principles of the empirical world those opposites of cold and hot, heavy and light etc. which, being attached to corporeal things, have no substance of their own.

Whilst these remarks take Plato himself for their target—and, at any rate the last, may seem to vindicate Anaxagoras (or perhaps Archelaus) against him—it is nevertheless plain that their immediate aim was to prevent his successors in the Academy, Speusippus and Xenocrates, from dogmatizing their old master's views.[2] Here then the question of the change to more than one ἀρχή in *Metaphysics* x.2 comes to the fore.[3] The purpose of emphasizing the existence of more than one ἀρχή can be twofold: it can be that the philosopher has made a metaphysical decision in favour of polytheism rather than monotheism, or else he may have made a physical decision in favour of a cosmos which is formed of basically different components, rather than of one identical substance

[1] Aristotle, *De gener.* II.1, 329a, 5 f. to the end of the chapter. I differ from F. Solmsen, *Aristotle's system of the phys. world*, Cornell Stud. XXXIII, 1960, 325, saying, 'what greater credit could he give to Plato etc.', because the passage is plainly polemical.

[2] As regards Speusippus, it seems clear that § 3 is directed against him, cf. the polemics in *Metaph.* XII.7, 1072b, 30 f., whilst § 5 goes against Xenocrates, cf. frgg. 42, 44, Heinze.

[3] G. S. Claghorn, *Aristotle's criticism of Plato's Timaeus*, 1954, 21, has based his remarks on a serious misapprehension when he states: 'Most of the Pre-Socratics had one or a small number of material bodies from which they thought the universe originated. These bodies were called στοιχεῖα or ἀρχαί (sing. ἀρχή), according to Aristotle, *Metaph.* 1.3, 983b, 7–11.' The fact is that Kranz's index does not contain a single instance for an identification of ἀρχή with στοιχεῖον, neither is there any reference to such an identification in the *Index Graecitatis Platonicae* by T. Mitchell, Oxford, 1832, which, however, is incomplete. It is also doubtful whether στοιχεῖον was not originally used to denote one of the 'four elements', and whether this was perhaps the reason why Aristotle, loc. cit., used both words, stressing their difference rather than their identity.

branching out in apparently varying but fundamentally identical forms. In the course of his philosophical evolution Aristotle came to the conclusion that both these approaches were necessary for a satisfactory understanding of the empirical world, and attacked the Pre-Socratics because of their concentrating upon the physical side of the problem. The passage in which this attack is delivered[1] does not yet belong to the last phase of his philosophical development.[2] When he wrote it he saw *ἀρχή* in the closest connection with *αἰτία*, principle and cause, but was perhaps more concerned with 'causes' than with an ultimate principle or cause. At any rate he was fully conscious of the fact that whilst the ultimate principle or cause was a purely metaphysical problem, there could only be causes in the plural in the physical world. The sharpness of his attack appears most clearly when he ridicules these natural philosophers as 'being by no means dissatisfied with themselves' at reducing the whole of nature to one principle, which could not be anything else than the One.[3] Here, therefore, a point is reached where the change from the singular *ἀρχή* in § 1 of our chapter *Metaphysics* x.2, to the plural *ἀρχαί* in its subsequent paragraphs receives a definite meaning. For here it becomes clear that within the realm of nature there must be at least two such 'beginnings', one active and the other passive, a 'beginning' in movement and a beginning in substance. The claim made in § 1 that the search for one ultimate 'beginning' of the empirical world, if unsuccessful so far, was nevertheless justified, is abandoned. In its place Aristotle put the claim—and this shows that we are dealing with an intermediate phase of his philosophy—that the passive 'beginning' could and should be proved by the examination of the coming into being of a phenomenon and its subsequent dissolution. This would show e.g. that timber was the passive principle from which a chair was made, and into which it was subsequently broken up.

This 'time factor', combining the 'beginning' and the 'ending' was abandoned, as we have seen,[4] during the final phase of Aristotle's philosophy. He no longer regarded it as a true touchstone. At the time, however, it seemed to be of service for the enquiry into the first principles of nature. The usefulness of such a combination of 'beginning' and ending appeared in particular in Aristotle's scrutiny of Parmenides,

[1] Aristotle, *Metaph.* I.3, 983b, 1 f.

[2] It is closely related to the discussion about form and matter in *Metaph.* IV.18, 1028a, 18 f., cf. W. Marx, *The meaning of Aristotle's ontology*, 1954, 48 n.53.

[3] Aristotle, *Metaph.* I.3, 984a, 27 f.

[4] Cf. *supra*, p. 112, n.3.

that representative of the earlier 'natural' philosophy who had had the greatest influence upon Plato.[1] Aristotle's criticism of this predecessor has a fundamental character. If he admitted that Parmenides felt rather than deduced the necessity of the two ἀρχαί of which we have spoken, he nevertheless asserted that the philosopher from Elea had been wrong already from the start. 'Nature' as it is met by humanity is not immovable but is actually in motion, at least partly.[2] This very change of motion and rest defies the theories of Parmenides and Melissus, just as it renders unacceptable those of the other natural philosophers that there should be only one physical ἀρχή.[3] Aristotle proclaims boldly first that there have to be more than one of them, and secondly that their number must be limited; and in the first book of his *Physics* he sets out to determine the finite number of ἀρχαί which the philosopher has to acknowledge as such. At the bottom of this criticism there lies the recognition that without contrast there cannot be any observation, so that motion without rest would be unobservable and, when transferred to the field of theoretical consideration, meaningless. Motion and rest are thus mutually dependent ideas within the context of 'nature'; and they show that the concept of one single, 'natural' supreme principle, as the 'natural' philosophers tried to establish, was equally meaningless. This thesis, which is elaborated at the start of the *Metaphysics*,[4] depended of course upon the identification of principle and cause, a fact which Aristotle does not seem ever to have realized consciously.

It has to be stated, however, that all this enquiry into 'natural' principles and causes could never satisfy Aristotle's constantly recurring demand that there had to be one, and only one, metaphysical 'beginning'. For the 'natural' principles were, by their very nature, subject to the law of rising and perishing.[5] Some eternal 'beginning' was demanded because of the eternity of the cosmos. Already at the time of his defec-

[1] H. Weiss, *Kausalität und Zufall*, 1952, 19. W. Marx, op. cit., 12 f., makes no contribution to our problem.

[2] Aristotle, *Phys.* I.2, 185a, 12. H. Weiss, op. cit., 19 n.4, suggests that the ἢ ἔνια (at least partly) in this passage may contain a reference to the 'unmoved mover'. This seems doubtful. It is unlikely, I think, that Aristotle subsumed this conception of the deity under some superior idea of 'nature'. Moreover, the fact that Aristotle assumed the existence of non-moving parts within nature is evident from his polemics with Plato, cf. H. Cherniss, *Aristotle's criticism of Plato*, I, 1944, 306. Systematically the contrast of moving and immobile is to be joined to his scheme of opposites, as hot and cold and moist and dry.

[3] Aristotle, *Phys.* I.1, 184b, 15 f.

[4] Aristotle, *Metaph.* I.3, 984a, 17 f.

[5] Aristotle, *Metaph.* V.3, 1027a, 29 f.

tion from the Academy, Aristotle put this demand into words in his *De philosophia*:[1]

The 'beginning' is either one or many. And if one, we hold what we have enquired for; if many, they are either orderly or in disorder. If in disorder, that which springs from them must be in greater disorder still, and there would be no cosmos, but shame (ἀκοσμία); and the unnatural (τὸ παρὰ φύσιν) would exist, whereas the natural (τὸ κατὰ φύσιν) would be non-existent. If they are orderly they either have been put in order by themselves or by an outside cause. But if they have been put in order by themselves, they have something in common which holds them together, and that is the ἀρχή.

This exposition shows clear traces of kinship with § 1 of *Metaphysics* x.2, which need not to be enumerated. For the comparison with § 1 makes it equally clear that Aristotle, by the time he wrote it, had become critical of the whole approach which he had outlined in *De philosophia*. For however close the sentence from § 1 may be ('how could there be order if there were nothing eternal, separate, and everlasting'), the statement from *De philosophia* fails to make it clear that that which is common to the 'natural' ἀρχαί, must itself be metaphysical.[2] It is also necessary to stress the fact that the fragment from *De philosophia* contains a clear reference to Plato's concept of 'disordered motion'; but it does not indicate how this concept was to be overcome, because it does not clear up the character and effect of these secondary ἀρχαί, with which the philosopher is dealing. In short, it appears that, if we were in possession of the context, the fragment would provide us with a still earlier stage of Aristotle's doctrine of the 'beginning'. As it stands, it provides us with a tantalizing uncertainty.

We can say no more than that the change between ἀρχή and ἀρχαί in *Metaphysics* x.2 indicates that the plural was meant to characterize these 'beginnings' as inside the empirical world, as 'natural'.[3] Here Aristotle was at the threshold of his gigantic work of separating physics and metaphysics. It is significant that he introduced his 'book of definitions', *Metaphysics* IV, with a detailed scrutiny of the various meanings of

[1] Aristotle, *frg.* 17, ed. Rose, 32.

[2] It has to be admitted, however, that the Christian scholiast, who preserved *frg.* 17 for us, has apparently omitted the closing sentence of the corollary, that which dealt with ὑπὸ ἔξωθεν, so that the conclusion which we have drawn here, cannot be regarded as being entirely beyond doubt.

[3] F. Solmsen, *Aristotle's system of the physical world*, Cornell Studies XXXIII, 1960, 92, 'to make physics in Aristotle's sense of the word possible, nature itself must be brought down to the physical level'.

ἀρχή, not only as showing the end of the road of earlier philosophy, but also as the opening of his new road.[1]

The enumeration of these meanings here is somewhat loosely connected and does not claim completeness; but it is sufficient for leading up to the closing statement: 'For all causes are *ἀρχαί*.' It has to be asked, however, whether Aristotle intended by this to maintain also that all *ἀρχαί* were causes, and that the two terms were synonymous? If so, this was an important advance beyond *Metaphysics* x.2, for it seems evident that he was, to say the least, not certain about this view when dealing with the mathematical *ἀρχαί* or with 'substance' in its later paragraphs. For he never refers to it there. He took a much more tortuous way when he refuted the claims of the One, the dyad, the point, and being to be regarded as *ἀρχή*. We have seen him, however, granting at a later time that claim to 'matter'; and it is of importance in this connection that he entered upon the question of its being a cause in two passages of Metaphysics VI. At the beginning of its chapter eight we are told that it is not matter as such which produces the bronze which in its turn produces a bronze ball, but rather the combination of material and form by a creative activity, which causes the effect.[2] Aristotle, therefore, insists that it is neither the material, nor the abstract, spheric form which are the immediate cause for the bronze ball to come into being, but creative activity. Nevertheless, he emphasizes somewhat later, in chapter seventeen, that 'substance' (i.e. material and form) is a 'beginning and some kind of a cause'.[3]

Since these considerations are still dependent upon Plato's *Timaeus*, and not in conflict with it, they are presumably not much later than *Metaphysics* x.2. I believe that they grew out of the contrast established there in § 1 between potentiality and actuality. However, with this new appreciation of 'substance' Aristotle abandoned the reason given there for his contention that its mere potentiality must prevent the philosopher from acknowledging that matter as such could be the *ἀρχή*.[4]

[1] Aristotle, *Metaph.* IV.1, 1012b, 34–1013a, 17.

[2] Aristotle, *Metaph.* VI.8, 1033a, 24 f. Aristotle is in this matter quite close to a view expressed by Plato, *Tim.* 50A. F. Solmsen, *op. cit.*, 122, remarks on this: 'It has been said that "in this point Plato's illustration is inadequate"; curiously enough, it would be quite adequate for Aristotle's concept of matter.' I agree with Solmsen's findings, but wonder about the significance of 'curiously'.

[3] Aristotle, *Metaph.* VI.17, 1041a, 16 f.

[4] It seems to me as if this statement has to be seen as very close to that other 'extreme statement' that matter might be identified with non-being, characterized as chaos in *De philosophia*. Cf. F. Solmsen, op. cit., 121 n.15.

His contention was now that in the sense of *conditio sine qua non* substance was a cause for the coming into being of the bronze ball. It is improbable that Aristotle should have adopted this view if he had not made two preliminary philosophical decisions: the one, that matter as such had an existence, if a metaphysical one; the other, that *ἀρχή* and *αἰτία* must be identical. For by understanding substance as a passive *cause*, he had to reconsider what from now on was no longer cause as such, but the active cause. The creation of the bronze ball was no longer to be considered as a form-*giving* to something which already existed in theory, but rather a privation (*στέρησις*) for the previously existing substance. From the metaphysical point of view this was a dangerous conclusion to draw. For it could well lead back to that position which, as we have seen, caused Anaximander and Empedocles, and—still more important—the Orphic movement, to give the place at the 'beginning' of creation to Nemesis–Adrasteia, or a similar avenging deity. And Aristotle was well aware of this.[1] It was probably the reason why he even treated the 'beginning' of motion—in contrast to Plato[2]—not as an ultimate, but as a secondary, 'natural' cause.

In *Metaphysics* II, the book of *aporiae*, it is true, Aristotle put the question 'whether the *ἀρχαί* were universal or separately directed upon each individual phenomenon, whether they were potential or actual, and whether they were concerned with anything else or merely with motion', and stated that that was very hard to answer.[3] This might be taken as an argument that his statement in *Metaphysics* VI.17, which we have just discussed, was of a later date. It is, however, evident that this question was intended as a polemical remark with direct reference to Plato's doctrine that the 'beginning' of motion was the ultimate principle of the empirical cosmos.[4] It is, therefore, probable that there was no great time lag between the two. Both seem likely to have been formulated before the last phase in the evolution of Aristotle's philosophy, i.e. before he had worked out the theory of the 'unmoved mover'. As regards these questions from *Metaphysics* II.1, it is evident that even at that time Aristotle did not regard them as unanswerable. With the help

[1] He refers to Orphic cosmogonic myths in *Metaph.* I.3, 983b, 27 f., cf. *supra*, pp. 32 ff. and 124, n.4.

[2] The myth in *Phaedrus* is probably as much under obligation to Orphism as is the *Timaeus*, cf. W. K. C. Guthrie, *Orpheus*, 2nd ed., 1952, 240.

[3] Aristotle, *Metaph.* II.1, 996a, 5 f.

[4] Cf. F. Solmsen, op. cit., 93, about Aristotle's determination to do without Plato's concept of the soul. To the literature quoted ibid. n.3, about the relation between nature and the world soul, add J. Moreau, *L'âme du monde*, 1939, 114 f.

of his doctrine of categories,[1] he came to assign to the 'cause of motion' a place among the four big causes 'from the beginning', essence, matter, motion, and purpose, describing it by the almost Platonic term ἀρχὴ κινητική.[2] As such he aligns it with the chief characteristic which he gives to nature, that it is self-moving;[3] but he relegates it at the same time to the rank of the secondary causes, those which exist in the realm of potentiality, whilst it was actuality (ἐνέργεια) to which he gave the first place.[4]

It is evident that this solution opened the way to that mature period of Aristotle's philosophy when he established the principle of the 'unmoved mover' as the 'beginning' of all causes. The four great causes were no longer supreme causes, still less were the 'beginnings' of nature, those opposites of which we have spoken, ultimate principles.[5] The solution which Aristotle now evolved meant, of course, the rejection of the mythology of Plato's *Timaeus*; but the way in which it was stated is on the whole free from polemical acerbity. Aristotle seems to have been satisfied that this last answer to the problem of the 'beginning' was the 'scientific' solution, and as such in the field of rational philosophy superior to Plato's mythology. Once again we are faced here with a development of Aristotelian thought, of which no more than the first

[1] M. Wundt, *Untersuchungen etc.*, 1953, 20 f., has rightly drawn attention to the importance of the essay *De categoriis* for the interpretation of Aristotle's *Metaphysics*. W. Jaeger, *Aristotle*, transl. R. Robinson, 1946, 46 n.1, has treated the essay as spurious, but stresses its Aristotelian character. In opposition to him—and not without recognizable bias—M. Wundt, supported by P. Gohlke, claims it as genuine. The reason given by them is, however, curiously deficient. They maintain that the essay is so unlike the mature philosophy of Aristotle that in later time nobody would have ascribed it to him without good historical reason. This is quite unconvincing since it presupposes that only experts made such ascriptions. As if the number of false ascriptions in Hellenistic times were not legion. In order to make such an argument valid, there must be at least one good reason in favour of the ascription. It seems to me, however, that such a reason can be adduced. I find it in the looseness of the collection of categories, which has its analogy in *Metaph.* II, the book of *aporiae*, as well as in *Metaph.* IV, the book of definitions. The aphoristic character of these three books seems to suggest that they originated in that period when Aristotle's 'first philosophy' was in the melting pot.

[2] Aristotle, *Metaph.* I.3, 983a, 24 f.

[3] Cf. R. G. Collingwood, *The Idea of Nature*, 1945, 81 (sub 7), 82 f.

[4] Aristotle, *Metaph.* VIII.8, 1049b, 4 f.

[5] It is undoubtedly correct to state with H. Weiss, *Kausalität und Zufall*, 1952, 28 f., 'the ἀρχαί of that which is by nature are several', but that does not mean that Aristotle ever abandoned his demand for one supreme principle. He only denied that such a principle could be found in 'nature'.

and the last stages can be fixed. The first I would try to find in § 6 of our basic quotation, *Metaphysics* x.2, where we meet with the confession rather than complaint that there are the greatest obstacles to an attempt at devising a purely scientific enquiry into the fundamentals as such. Fundamental enquiry, it is claimed there, has to deal principally with specific questions, although it leads in the end to fundamental principles by systematically reducing its object to basic data. The error of such specialists, however, when they attempt to establish fundamentals by the inductive method, is made clear in *Metaphysics* v. There Aristotle explains that scientific research cannot lead beyond the establishment of principles and causes of things that exist. Such principles and causes, therefore, are valid only with regard to the special kind of existence of the things which have been subjected to such scientific examination. It thus reduces specific phenomena to their simplest and most precise causes, but it is unable to draw any conclusions about their reality (*Dasein*), let alone existence in general.[1] Such a method, therefore, cannot provide any final certainty since it relies primarily upon observation without establishing the reality of the things observed, or analysing the hypothesis which forms the basis of the experiment observed. In view of the fact that it is medicine and mathematics which are the sciences criticized, it is probably correct to suggest that the opponents attacked by Aristotle in this way were members of the early Academy with their leanings towards Pythagoreanism.

However that may be, the important thing is that Aristotle now set out to establish an ontology as part of his philosophy. If it is generally true that Pre-Socratic philosophy had dealt entirely with the ontological problem then it would follow that these philosophers were unaware of the fact that this was only one philosophical discipline. Despite Plato's universality it is true to say that it is the third book of Aristotle's *Metaphysics* which marks in this respect the turning point of European philosophy. For Aristotle saw the problem in its separation as well as in its entirety, even if it had to be admitted that its definition was only achieved by the Christian schoolmen.[2] Whilst in *Metaphysics* x.2 he had still only pointed out the great difficulty of establishing a theory of mere being, as distinct from the being of certain phenomena or groups of phenomena, Aristotle resolutely started its third book with the statement:[3]

[1] Aristotle, *Metaph.* v.1, 1025b, 2 f.

[2] This point has been made strongly and rightly by W. Marx, *The meaning of Aristotle's ontology*, 1954, vii.

[3] Aristotle, *Metaph.* iii.1, 1003a, 21 f., cf. the translation by W. Marx, op. cit., 4.

There is a science which enquires into being quâ being, and into its properties as such. This science is in no way identical with the specialist sciences so-called. For none of these enquires unconditionally into being quâ being; but having separated a special aspect of it, they all contemplate the particular quality of that, as for instance the mathematical science does.

Here we find once more the polemics against the Pythagorean tendencies of the early Academy. 'For', Aristotle continues, 'as we seek for the principles and supreme causes, we have to presuppose that they are of necessity of a nature entirely their own.' This attack is, I believe, explained and enlarged by the next following sentence:

If, therefore, those who enquire into the elements of that which exists, had enquired into those principles then it would follow that the elements belonged to the whole of existence not incidentally (according to their specialist approach), but because of existence. Therefore it is up to us to grasp the first causes of existence quâ existing.

This passage, which may be slightly damaged,[1] has its ethos in Aristotle's strong protest against a philosophy which identified mathematics with ontology, i.e. the Academy under the leadership of Speusippus. Number, it says, may be a principle, even *the* principle of mathematics; but that does not show that it is a principle of being, and may not even show that it exists. The statement that number is a principle is merely accessory to the specialist approach of the mathematician. Evidently, it had not been so very long ago that the Academy had slammed in Aristotle's face its gate with the proud inscription: 'No admission for non-geometricians.'

It is, I think, evident that the analysis of 'being' in the first chapter of *Metaphysics* III was worked out fairly soon after Aristotle's attack upon Plato's successors in *Metaphysics* X.2, where he had criticized them in § 6 because they had not realized the independent character of ontology. On the other hand, the exposition of the manifold character of 'being' and of the One, which is to be found in *Metaphysics* III.2, shows great similarity to that given in Aristotle's *Physics*.[2] The difference between

[1] It is here that my translation and understanding of *Metaph.* III.1 differs from that of W. Marx, op. cit., 5. For one thing, I cannot take the changes between the plural and the singular of *τὸ ὄν* as seriously as he does; secondly, as in this instance we are faced with lecture notes, I venture to suggest that before the last sentence a thought has been omitted, whether it be by Aristotle himself or by one of his students; and finally, I regard the whole chapter as plainly polemical, whereas Marx has understood it as a general statement.

[2] Aristotle, *Phys.* I.2–3, 185a, 21 f.

the two passages consists mainly in the fact that in his *Physics* Aristotle openly polemicizes against Parmenides and the Eleatics, whereas in *Metaphysics* III.2 no such polemical edge is brought into the open. I cannot see any decisive reason for regarding either passage as being much earlier than the other. It is Miss Weiss who has based her valuable remarks on Aristotle's thesis, *πολλαχῶς λέγεται τὸ ὄν*, 'being' has many connotations, upon an appreciation of the passage from Aristotle's *Physics*.[1] It is perhaps for this reason that I feel that the statement in *Metaphysics* III.2 deserves our special attention, and that it should be regarded as the programme of the new discipline of ontology.[2] However, by and large Miss Weiss has arrived at similar conclusions to those which follow from the two statements in *Metaphysics* III.2: 'Being is discussed under several headings, but in the direction of Oneness, and of a sort of common nature', and consequently 'being' is discussed in various ways, but as altogether directed towards one principle (*ἀρχή*).[3] Already in these sentences it appears clearly that Aristotle is concerned with establishing an analogy between 'being' and 'the One';[4] and Miss Weiss has shown convincingly that this analogy is derived from the fact that these two conceptions, 'being' and 'the One', respond in the same way to Aristotle's method of dialectical analysis. For he regarded this as an observed fact—assuming that 'Aristotelian' logics were the only logics available to the mind—and from this he deduced that logical thought was necessarily directed towards the establishment of one supreme principle in the whole system of being. The logical impossibility of ever arriving in this way at the definition of the supreme principle of his philosophy, his doctrine of the godhead, was nevertheless recognized

[1] H. Weiss, *Kausalität und Zufall*, 1952, 31 f.

[2] The magic of numerical sequence is nowhere more dangerous than in interpreting Aristotle. Miss Weiss, I believe, has concentrated upon *Phys.* I. 2–3, mainly because her ultimate aim is the analysis of *Phys.* II.4–6. However, it cannot be shown, and Miss Weiss has not even attempted to show it, that Aristotle wrote *Phys.* I.2–3, before he wrote *Phys.* II.4–6; and it is thus doubtful whether Miss Weiss' own work has really benefited by her choice. More serious is the fact that W. Marx, op. cit., 4, following innumerable others, has treated *Metaph.* III.1, as 'the ideal formula of any ontology', whereas it is clearly an introductory tussle with Aristotle's opponents; and that he has thus deprived the real programme in chapter 2 of some of its significance, cf. esp. p. 6.

[3] Aristotle, *Metaph.* III.2, 1003a, 33 f.; 1003b, 5 f.

[4] It is remarkable, and I think a serious omission, that W. Marx, op. cit., 33 n.12, has left the *πρὸς ἕν* untranslated, and has not taken it into account in his analysis of 1003b, 5 f., either.

by Aristotle,[1] and as I believe, expressly stated in his treatises on logic.[2]

The way in which early Peripatetic philosophy and, as we shall see, probably Aristotle himself, tried to overcome this apparent stalemate is explained in the *elenchus* so-called of the first book of the *Metaphysics.* The very fact that this book is in all probability post-Aristotelian makes it valuable for our enquiry. For, although it may contain non-Aristotelian additions, its origin, presumably on the basis of lecture notes taken at the Lyceum,[3] entitles us to assume that it presents the last stage in the evolution of Aristotelian thought, but does not face its readers with any basic changes brought about by the later Peripatetics. Here we find one point emphasized which, unless I am mistaken, had not been mentioned in any considerations about *ἀρχή* which we have found in the earlier philosophers, including not only the extant fragments of the Pre-Socratics, but also Plato's *Timaeus* and even in Aristotle's genuine writings. It is that the ultimate 'beginning' must be closely related to ultimate truth. Of course, all the earlier philosophers had aimed at teaching truth, and separating it from falsehood. However, the new conception in the *elenchus* is that in the philosophical argument as such an intrinsic verity must be discernible, which rests upon its logical conclusiveness. This verity in its turn reflects that ultimate truth, which is the necessary concomitant of the ultimate *ἀρχή*:

For we do not know the truth without the cause. Everything, however, is most nearly defined by that, by which it retains its identity in its influence upon other things, as for instance fire by being essentially hot: for it is the cause of heat in the other things. Consequently the essentially true will be the cause for that which follows from it being true (also). Therefore it is necessary that the principles should be essentially true. They cannot be accidentally true, neither is there for them any further cause of their existence (beyond the ultimate 'beginning'); but they are rather that cause for everything else which, inasmuch as it has part in existence, so also has part in the truth.[4]

With this statement, and indeed with the whole argument which precedes it, which aims at showing that philosophy is the science of

[1] W. Marx, op. cit., 33 f., is correct when saying, 'that only the natureness of substantiality is accessible to man'; but if he had considered the logical analogy between 'being' and the One, he might not have missed the change from the singular *οὐσία* to the plural *οὐσιῶν* in his translation of *Metaph.* III. 2, 1003b, 17 f., and even have considered its significance.

[2] Cf. Aristotle, *Top.* IV,. 121b, 1–14; *Analyt. post.*, 72a, 6 f.

[3] Überweg–Prächter, *Philosophie d. Altert.*, 12th ed., 1926, 367 f.

[4] Aristotle, *Metaph.* I (elen.) 1, 993b, 23 f.

truth, a new basis for Greek philosophy, and indeed all human philosophy, was established. For this statement is meant to do two things. First it establishes that existence and truth are correlated so that there can be no existence without truth; and this in its turn has the effect of showing that the principles of things eternally existing must be eternally true. As a hypothesis (because it cannot be proved here) I propose that this statement contains a reference to the celestial phenomena, the spheres and the stars, and to their *archons*; and they are the guides to the ultimate truth. However that may be, the vehicle for the imparting of existence and truth to things temporarily existing is causality, and although this is not expressly stated, it is clearly implied that the method of testing this truth is logic. This brings us to the second achievement of this summary of Aristotle's philosophizing about the 'beginning', that in which the great divorce from earlier Greek philosophy is to be found: as all things existing exist by truth, so also does truth itself exist. There exists an objective, discoverable truth in the principles (*ἀρχαί*) of the things eternal.

In the course of these enquiries it has been stated repeatedly how from Anaximander onwards the energy of Greek philosophical thought had been towards the reconciliation of the mystery of a revealed religion with the newly observed facts, systematized by natural science as well as with the mathematical discoveries of pure reason. Such had been the ethos of Empedocles and of the Pythagoreans in particular. Parmenides and the Eleatics, despairing of the power of observation, had relied entirely upon the revelation described in the introduction of Parmenides' poem —and in this sense it had been his myths much more than his reasoning which had put Plato in such a close affinity to Parmenides. The belief in the existence of an objective truth had kept its position as a religious conviction; but the confident hope of the human mind being able to discover it, which had been so strong in the sixth century, had in the course of the fifth given way to despair. It is well known how the starting point of the rationalist, philosophical movement amongst the Ionians is found in Protagoras who had stated: 'About the gods, I know not whether they are or not; neither have I any knowledge as to what they are.'[1] He thus dismissed religious intuition as a source of the recognition of the truth. On the other hand, it was Anaxagoras, the very philosopher whose doctrine of the Nous showed him, said Aristotle, to be the only sober one amongst the intoxicated,[2] who had resigned

[1] Protagoras frg. B 4, Diels–Kranz, 5th ed., 1935, II, 265.

[2] Aristotle, *Metaph.* I.3, 984a, 15 = Anaxagoras frg. A 58, Diels–Kranz, 5th ed., 1935, II, 20.

himself that 'because of the weakness (of the senses) we are incapable of distinguishing the truth'.[1] He had thus abandoned the hope of discovering truth by way of scientific observation. However, his example became essential for the development of Aristotle's metaphysical doctrine of truth, because of the part which he had assigned in his cosmogony to the action of the Nous. For in this conception of the Nous the two essential properties demanded by Aristotle for his ultimate ἀρχή, truth and intelligibility, were combined.

The comparison of Anaxagoras' conception of the Nous with the pseudo-Aristotelian statement just quoted is legitimate, even though pseudo-Aristotle speaks of eternal ἀρχαί in the plural. It may be granted that attention should be paid to a recent attempt at showing that the change from 'substances' to 'substance' in *Metaphysics* III.2, does not yet point to an ultimate ἀρχή, but only to the fact that Aristotle regarded 'substance' as an ἀρχή,[2] it is nevertheless evident that in chapter 2 of the *elenchus* an ultimate ἀρχή is envisaged, which has to be the one supreme cause of existence, and thus can be established by way of logical deduction. Here the author, continuing the statement which we have just discussed, claims:[3]

That there is a supreme principle (ἀρχή τις), and not just an unlimited number of causes for the things which exist, be it in an infinite series or in limitlessness as such,[4] is clear. For neither can the coming of one thing out of another, as it were from matter, continue indefinitely, as flesh of earth, earth of air, air of fire, and so on without stop; nor is that admissible with regard to the beginning of motion, as man being moved by air, that by the sun, the sun by strife which itself is limitless.

For if such lists were to continue *ad infinitum* they themselves would become unintelligible. For it is the very principle of intelligibility, which prohibits both the progress and the regress into the limitless: 'For those

[1] Anaxagoras frg. B 21, Diels–Kranz, 5th ed., 1935, II, 43. It is to be regretted that R. Bultmann, *Theol. Wb.* I, 1933, 239 n.20, has used this fragment solely for the illustration of the grammatical usage of κρίνειν τἀληθές. The philosophical scepticism of classical sophism, and its importance for the formation of the idea of truth in pre-Christian Greek thought, so significantly expressed in John 18:38, he has left untouched.

[2] W. Marx, *The meaning of Aristotle's ontology*, 1954, 35 f.

[3] Aristotle, *Metaph.* I. (elen.) 2, 994a, 1 f. This passage has to be understood in the light of *De anima* III.6, 430b, 5, 'for it is the Nous that is the unifying force in all this', cf. J. Vahlen in F. Trendelenburg, *Aristotelis De anima*, 2nd ed., 1877, 415 n.

[4] The two expressions used here refer to the 'two ways of coming into being', procreation and evolution, discussed shortly afterwards, 994a, 22 f.

who deny this abolish all understanding, for true knowledge is only established by going back to the fundamentals (τὰ ἄτομα).'[1]

All this is stated here not so much by argument and logical deduction as by assertion. Any reader of the *elenchus* will feel clearly that the real spadework, by which these views had been established, had been done elsewhere. It is the reference to the 'beginning of motion' in particular which leads to an examination of Aristotle's *Physics* for the counter-checking of these views.[2] There Aristotle remarks:[3]

There are, therefore, three lines of research, the first about the immobile, the second about the moving but imperishable, the third about the perishable, so that the reason why should become clear to whoever enquires into matter, and into existence, and into the first moving power.

It seems very probable that this was the programme which was in the mind of the writer of the sentences quoted from the *elenchus*, which we have just discussed. They are at the same time also dependent upon the —presumably final[4]—statement about ἀρχή which is to be found in the eleventh book of the *Metaphysics:*[5]

For the principle, and that which is first of the things existing, is immovable in itself as well as in its relation to the outside, yet it sets into motion the primary, eternal and one, motion.

Whatever may have been the time when this thesis was applied to the astral religion outlined in the later part of the same chapter,[6] the thesis itself clearly reflects the programmatic statement established at the outset of the *Physics*, and quoted by us just now.[7] Between these two fixed points, the conclusions which we have drawn from the *elenchus* should appear clearly as genuinely Aristotelian.

[1] Ibid., 994b, 20 f.

[2] The fact that the *elenchus* should be seen in close connection with the *Physics*, has been established long ago by W. Jaeger, cf. Überweg–Prächter, *Die Philosophie d. Altert.*, 12th ed., 1926, 368.

[3] Aristotle, *Phys.* I.7, 198a, 29 f.

[4] W. Jaeger, *Aristotle*, transl. R. Robinson, 2nd ed., 1948, 342 f.

[5] Aristotle, *Metaph.* XI.8, 1073a, 23 f.

[6] Such a connection has appeared to us also in *Metaph.* I (elen.) 1, 993b, 23 f., cf. *supra*, p. 132.

[7] The close affinity between these two passages may be seen when the sentence from *Metaph.* XI.8 is brought into relation with the second point of Aristotle's programme in *Phys.* I.7. For there he continues to discuss 'the unnatural beginning' of motion, 'not having in itself the beginning of motion', *Phys.* I.7, 198b, 1, and in *Metaph.* XI.8, 1073a, 27, he adds his remark on the *πρῶτον κινοῦν, ἀκίνητον καθ' αὑτό.*

Here then a way was found to resolve the misgivings of Pre-Socratic scepticism. If the existence of the empirical world was based upon a logical truth, then the deception of the senses which endangered the apperception of the empirical world could be checked with the help of reason. Those misapprehensions which occurred were only incidental since it was established that causality and logic were dialectically connected by the immanent truth which, descending from an absolute truth, established the phenomena of the empirical world which were caused by ultimate 'beginnings', and finally an infinite 'beginning'. It is easy to outline this conception on the basis of the work done by Aristotle. He himself was temperamentally a conservative thinker, and hesitant to reject conclusions which in the field of speculative philosophy had served well to elucidate its problems. It was thus by way of modification rather than of radical change that Aristotle approached Plato's doctrine of the world-soul. We know from Cicero[1] that Aristotle in his early dialogue *De philosophia* had not in this respect differed at all from the teaching of his master. It also appears from Aristotle's known works that even in his last period, after he had established his theory of the 'unmoved mover', he still found a place for the 'self-moving mover' of Plato in his cosmology. For we find in his *On coming into being* the statement:

> The moving agent is to be understood in two ways. For where there is the beginning of motion the same appears to be moving (for the beginning is the first of the causes) as well as also the last in the direction of receiving motion, and coming into being.[2]

It appears from this that Aristotle was still convinced that the supreme cause, in order to set the causal nexus into motion, had to move itself. This, however, did not demand that it should move without ceasing: 'For nothing prevents the first moving agent from being unmoved whilst the motion continues; at times this is even inevitable.'[3] And it is

[1] Cicero, *De nat. deor.* I.33 f., saw Aristotle's early astral theology in close connection with that of Xenocrates and, as J. Moreau, *L'âme du monde*, 1939, 120 f., has shown, also with that of the ps. Platonic 'Epinomis'. This statement is valid for *De philosophia* Book III (Aristotle frg. 26, Rose); but I venture to express some doubt whether scholars from G. F. Schoemann, *M. Tulli Ciceronis De nat. deor.*, 2nd ed., 1857, 50 f., down to J. Moreau, loc. cit., have done wisely when drawing conclusions from Aristotle, *Metaph.* XI. 6 f.; *De caelo* I.2; II.1 & 3; *Meteor.* I.3, upon the contents of *De philosophia* Book III.

[2] Aristotle, *De gener.* I.7, 324a, 25 f.

[3] Ibid., 324a, 30 f. Cf. W. J. Verdenius and J. H. Waszink, *Aristotle on coming-to-be*, 1946, 43 f., defending successfully the genuineness of the words πρῶτον κινοῦν

to be noticed that the point which is here made by Aristotle concerns the empirical world in which, so he maintains, there is no difference in meaning, whether the cause of motion is found in the original cause of all motion or in the immediate cause of one individual moving body.

This conclusion arrived at in *De generatione*, seems to be echoing a far more general statement made in Aristotle's *Physics*, which indicates that he firmly placed the Platonic world-soul within the system of the empirical world. In this work he saw two 'beginnings' of natural motion, the one immanent in nature, the other not,

> because it does not contain in itself the principle of motion. For such is the case if an immobile entity imparts motion, as for instance the universally immobile, and the first of all things being, and the individualizing agent (τὸ τί ἐστι), and the form. For that is the supreme power (τέλος) and the ultimate cause. Therefore, since nature follows the law of cause and effect, it is necessary to be also aware of this, and at each step an account of the cause has to be laid by stating: Since A is so, therefore B has to be so.[1]

Aristotle, therefore, makes the claim in this statement that a certain cause A, if taken separately, can of necessity have only one effect B, since it would be impossible to understand nature if this thesis were to suffer any exception; and since the intelligibility of nature is the necessary corollary to the fact that its first principle is truth. From this it follows that the law of cause and effect itself is also true because all causes are ultimately resolved in one supreme cause, the unmoved mover.

The unmoved mover is thus established as existing, because the very conception of cause and effect would be meaningless, i.e. removed from any concept of rational truth, if it referred to an endless series of events. He is supreme as well as 'infinite',[2] and we may describe him as god. Natural science, therefore, which in a very special sense makes the infinite the supreme 'beginning',[3] is therefore in the last resort an integral part of theology in Aristotle's philosophical system. In this sense he could validly maintain that there was no difference in meaning when the

---

against the reading of a simple πρῶτον. Whilst our argument would not be invalidated if this reading were accepted the recommendation of these two scholars supports it more strongly.

[1] Aristotle, *Phys.* I.7, 198b, 1 f., cf. W. Marx, *The meaning of Aristotle's ontology*, 1954, 57 f.

[2] Cf. A. Edel, *Aristotle's theory of the infinite*, 1934, 85, where the author analyses the conception of 'infinite regress', and his very persuasive defence of an 'actual infinite' as the summit for Aristotle's system, ibid., 92 f.

[3] Cf. P. Seligman, *The 'Apeiron' of Anaximander*, 1962, 57 f., with its criticism of Aristotle, *Phys.*, 203b, 4.

term 'cause of motion' was referred to either the ultimate source of all motion, the supreme 'beginning', or to the immediate cause of an individual motion.[1] His firm belief is that the physical and the metaphysical are inseparable. However, it has to be said that all this was logically proved only in so far as motion as such was concerned, but neither with regard to the objects which are set in motion nor their ways and their aims; and the question as to the meaning of 'motion as such' may have been altogether alien to Aristotle's mind. However, Aristotle did conceive of a motion beyond the moving of material bodies. This may be seen from his attempt at explaining the movements of the celestial bodies, where he has given a classification of these movements, which is of a special importance for his conception of the ultimate 'beginning':[2]

For it is the best by far for all (the celestial bodies) that they reach the perfection described; but if not, it is progressively better the nearer the best is approached. And for this reason the earth is not moving at all, and its nearest neighbours only slightly. For these do not even arrive at the last stage, but only to such a one that they may just reach out to the supremely divine beginning (θειοτάτη ἀρχή). The first heaven, however, reaches it in one single movement; and the celestial bodies between the first and the last reach it, but reach it only by way of numerous motions.

This astral rather than astronomical conception not only shows how in Aristotle's thought the progressive refinement of the moving matter in the higher spheres eases the obstacles which prevent the contact between them and the source or supreme 'beginning' of all motion, the unmoved mover, but it also guides us to the second 'beginning' in Aristotle's thought.[3] This Aristotelian dualism is still strongly influenced by later

[1] Aristotle, *De gener.* I.7, 324a, 25 f. The slight variations in the interpretation of this passage discussed by Verdenius–Waszink, *Aristotle on coming-to-be*, 1946, 43 f., do not touch the point we have made here.

[2] Aristotle, *De caelo* II.12, 292b, 16 f. The close connection between *De caelo* II.9–12, and *Metaph.* XII.8, should be, I believe, self-evident. It has also been stated from the days of Simplicius' commentary on *De caelo*. Now, whilst it is an undoubted achievement of W. Jaeger that he has proved the coherence of *Metaph.* XII.8, against all who have tried to separate its astronomical passage, it is unconvincing when he (*Aristotle*, transl. R. Robinson 1948, 348 n.2) concludes that its connection with the passage from *De caelo* should no longer be maintained. For this passage presupposes the doctrine of the unmoved mover and, therefore, is not to be ascribed to an early period of Aristotle's philosophical development, as W. Jaeger's thesis implies.

[3] Aristotle, *Meteor.* I.9, 346b, 16 f., says, *κυρία καὶ πρώτη τῶν ἀρχῶν ὁ κύκλος ἐστίν*, and this saying strongly tempts the reader to assume that Aristotle when he wrote it had not yet evolved the theory of the unmoved mover. W. Jaeger, cf. the

Platonic thought—or mysticism. This recognition is forced upon us when we consider the following exposition in the eighth book of the *Physics*:[1]

Since the preceding argument should be evident to all, we now will endeavour to show the respective nature of each of the two, namely that the one is eternally immobile, and the other eternally in motion. Proceeding from there, and claiming that everything moving is moved by some agent, and that such an agent must itself either be unmoved or moving, and if moving either must be moved by itself or by something else eternally, we have reached the conclusion that the beginning of things moving, in so far as they are moving, is the self-moving mover; but of the universe as a whole, the immobile. We also have an evident instance for the existence of self-moving movers in the whole of the animated as well as animal life.[2]

Clearly, Aristotle shows in this passage the same certainty with regard to his doctrine of the 'unmoved mover' as he does in *Metaphysics* XII.8. However, he has found here that his doctrine is compatible with Plato's great conception of the cosmic soul.[3] There can be little doubt that this doctrine had a profound influence upon Aristotle's thought. He employed it not only for the explanation of the movements of the celestial bodies, and the 'life' of the animals, but also as the basis for the explanation of physical phenomena in the sublunary sphere; and it is here that we find at last the justification for Aristotle's using the words ἀρχαί and στοιχεῖα as synonyms:

---

preceding note, even appears inclined to yield to this temptation. That would then mean that we would find here an almost unchanged Platonism in one of Aristotle's esoteric writings, which would be highly exceptional. It seems therefore wisest to adopt the resignation of F. Solmsen, *Aristotle's system of the physical world*, Cornell Studies XXXIII, 1960, 396, 'that Aristotle here does not aim at more than providing valid and intrinsically coherent explanations of the whole range of meteorological phenomena', and to refrain from the drawing out of any lines that may seem to be indicated. Aristotle has said no more in this passage than that the supreme sphere in heaven is, as its first 'beginning', still within the cosmic system, and not the principle of the whole cosmos.

[1] Aristotle, *Phys.* VIII.6, 259a, 27 f. Cf. F. Solmsen, op. cit., 226, 'that eternal movement was numerically one was settled when Plato had identified self-mover, first mover, and eternal mover and associated this principle with the never-changing celestial movements'.

[2] I feel sure that the contrast between ἔμψυχα and ζῷα, made here, refers to that between the supra-lunar sphere of 'animated bodies', and the sub-lunar sphere of 'animals'.

[3] The passage appears to give a summary of Aristotle's conflict with Speusippus, as outlined by J. Moreau, *L'âme du monde*, 1939, 131 f.

The beginning

We have previously defined that there is one ἀρχή for the bodies of which the nature of the bodies carried in the circular movements (of the heavenly spheres) consists; but we now say that there are four bodies because of the four elements (ἀρχαί), and that they have a twofold motion, the one away from the centre and the other towards the centre. They are four, fire and air and water and earth; and of these the highest is fire, and their lowest basis is earth. Two of those show an analogous condition to the one of the other two respectively. That means that air is nearest to fire, water to earth. The entire cosmos, however, surrounding the earth, consists of the said bodies. Of this (the cosmos) we say that it is the aim of our investigation. For this cosmos is of necessity continuous with the superior reaches in so much that all its potentiality is ruled from there. For from where the beginning of motion comes to the universe, there the first cause is to be found. Furthermore, that (beginning) is eternal and has no end of motion in space, but is for ever one with the end; but it has arranged all those bodies (the elements) in circumscribed regions, each separate from the other. Therefore it is necessary to regard as the causes of all that which happens concerning the cosmos, the fire, the earth, and their allies in the part of matter—for thus we call the underlying and passive cause—but that which is causal in the manner of the beginning of motion, it is necessary to regard as the force of all that which is ever moving.[1]

This passage seems to remove the last and most important difficulty in Aristotle's conception of the 'beginning' of the universe, by replacing the abstraction of 'matter', which can be neither abstract nor concrete because abstract matter is a contradiction in itself, whilst concrete matter is non-existent because it cannot be reached by scientific observation. It is, however, true to say that the recognition embodied in it cannot be shown to have influenced his cosmology to any large extent.

[1] Aristotle, *Meteor.* I.2, 339a, 11 f.

PART TWO

The ‘beginning’ in the philosophy of creation during the Hellenistic period until the Fourth Gospel

8 Summary of the earlier development

With the passage from Aristotle's introduction to his *Meteorologica* just quoted, which opens up vast areas of astrological speculations which could be and were logically derived from his speculations on the 'beginning', our scrutiny of the sources for this part of his philosophy is completed. It remains for us to sum up not only his views but indeed the whole evolution of Greek classical thought on the 'beginning'. For at the time of his death, 322/1 B.C., classical Hellas had come to an end. Time and time again it had been shown that its city-states were not equal to the demands of a new era. Thebes lay in ashes, Sparta's power was no more, and Athens had suffered terribly at Chaeronea and in the subsequent upheavals, especially since the death of Alexander. It is a miracle that a genius like Aristotle should have arisen at this time, not only to complete the circle of classical philosophy, and to save of the glories of earlier Greek thought as he did save, but also to indicate and even lay down what its future trend would have to be.

For these were indeed Aristotle's achievements. Thanks to his collections we have been able to follow the road which started with Anaximander, who first took up the question of the 'beginning' as a purely noetic problem. He was certainly not unmindful of the existence of physical problems; but was aware of the fact that the connection between the unchangeable infinite and the constantly changing phenomena of the empirical world was not to be found in the field of physical science, but lay beyond it. He by no means abandoned physical research; but he grasped the fact that it had to be based upon transcendental axioms, of which the most significant in his eyes was that of equal justice.[1] It seems that he was inspired to do so by the demand for a theodicy, and that his statement about the 'beginning' was founded more particularly upon the mystical philosophy of the Orphics, and their cosmogonies, than upon Thales' natural philosophy. This line was taken up rather by Anaximander's pupil Anaximenes. The noetic problem, however, was not especially a 'Milesian' problem, as is shown by the fact that it was

[1] Cf. on the nature of this axiom P. Seligman, *The 'Apeiron' of Anaximander*, 1962, 71 f. On the reappearance of this idea in Aristotle cf. Ch. H. Kahn, *Anaximander*, 1960, 186 f.

taken up and restated by Pythagoras and his school. It appears clearly, however, that they removed justice from the place of leading axiom, and replaced it by pure, mathematical reason. For they found the 'beginning' in number. Here was progress: for the rules of pure reason are discoverable by the examination of numbers and geometrical figures, whereas the rules of natural justice can never be stated incontrovertibly. At the same time the Pythagoreans lost by their subjection of justice to the supreme rule of mathematical reason to a great extent the contact with the physical world, and in particular that with Thales' rule of the equilibrium of physical forces. Still greater was their loss of contact with the conception of a social community of gods and men; and it was on this point that Empedocles parted with Pythagoreanism. For the method of number symbolism, employed by the Pythagoreans in the field of ethical philosophy, proved to be arbitrary.

At the same time the natural philosophers, observing that the empirical world was constantly changing, put in question the value of the noetic approach to the understanding of the world on the basis of a logical principle, a 'beginning'. For such an approach omitted change, the most characteristic quality of this world, in its considerations. It is not surprising that Heraclitus, the first to propose the theory of universal motion,[1] avoided the use of the word *ἀρχή*. He aroused such a vivid discussion with this attitude that over a century later Plato still described the two parties to the conflict by their nicknames, as the 'runners' and the 'stayers'.[2] It has also to be stated that Empedocles' attempt at a compromise by describing the two essentially different groups of conceptions, love and strife on the one hand and the four elements (which he apparently introduced into the philosophical discussion) on the other, without distinction as 'beginnings', did nothing to clear up the problem. It may, however, be regarded as a forerunner of Aristotle's later distinction between 'active' and 'passive' causes or 'beginnings'. However, even the 'runners' admitted the necessity of explaining the rush of physical phenomena surrounding man in the empirical world by reducing them to as limited a number of basic data as possible. Yet these data had to be immanent in the empirical world. From the time of Theophrastus the doxographers compiled lists of natural philosophers with the *ἀρχαί*

[1] Cf. Diels–Kranz, 5th ed., I, 141, 19, Heraclit. frg. A 1. I am fully aware of the fact that later doxographers, cf. Hippolytus in H. Diels, *Doxogr.*, 555, 5 f., ascribed the saying *πάντα ῥεῖ* to Thales; but I feel certain that Diels–Kranz have done right in not including this saying even amongst the A fragments of Thales.

[2] Plato, *Theaet.*, 181A, 1; *οἱ ῥέοντες* and *οἱ τοῦ ὅλου στασιῶται*.

proposed by them; but the term is a Peripatetic one, and was not used by the authorities quoted by those doxographers. The 'stayers', on the other hand, either denied the very existence of the empirical world, as Xeniades, claiming it to be created out of nothing—and nothing cannot be a 'beginning'; or, like the Eleatics, questioned its recognizability by sense apperception. It seems that Xeniades did not find any great hearing, although there are traces of his doctrine which show that he found an echo, especially in the Peripatetic school in the pre-Christian period.[1] The Eleatics, however, established a more powerful argument by showing that 'existence' is one, and decisively a noetic problem, which cannot be solved by scientific observation. Consequently they demanded that the universe must be one, that it must have as its essence eternally fixed noetic cosmos. The 'runners' in their turn, abandoning their earlier confidence that their basic data, which would prove to be the hinges of the empirical world, might be established by scientific observation, developed a theory of higher materialism—atomism.

It was the great merit of Anaxagoras that he, accepting the Eleatic demand for the unity of the universe, tried to apply it to the empirical world. He established two principles for it, the one, the Nous, as an active principle, the other the smallest particles of the combined matter, called by Aristotle the homoiomeres, as the passive principle. It has to be admitted that this attempt was still rather crude, being the first consciously teleological approach to the problem of the 'beginning'. It took into account that motion, change, was the chief characteristic of the empirical world, so that the 'beginning' appeared as the beginning of motion; but it demanded that the 'empirical' character of the world should be derived from the 'all-embracing' substance, which was static. It demanded that—in order to be immanent in the empirical world, as the principle of motion—the Nous must also consist of substance; but in order to safeguard its autonomy it had to be of such a substance which was wholly separate from the 'all-embracing' substance. Consequently the substance which contained the passive principle was not 'all-embracing', or else the Nous was not substance. Anaxagoras also seems to have never realized that neither his Nous nor his substance were observable phenomena. Nevertheless, his attempt was of the greatest significance. The important thing was that his deep concern, his power of observation, rested in the first place upon emotional foundations, and it was of minor importance if he could be proved logically wrong. In contrast to the

[1] Cf. Ps. Aristotle, *De X.Z.G.* III.1, 975a, 8 f., cf. my 'Creatio ex nihilo', *Studia Theologica* IV, 1950, 23 f.

abyss of pessimism with which he was surrounded by contemporary philosophers and poets,[1] he found that the joy of observing creation made life worth living. His claims were profounder than his proofs. For the sake of honesty—he said that he could have no rational notion whether there were gods or not, and was daubed 'the atheist' on this count by his opponents, the legalistic pessimists—he took the most consequential step of introducing the causal conception of the 'beginning', and with it the idea of creation in the sense of manufacture or production of the empirical world.

Anaxagoras' attempt was of such great significance because he alone, of all his fellow philosophers, realized that whilst the noetic attempt at making a valid pronouncement about the 'beginning' of the empirical world would of necessity lead to a theory of two worlds, it was this empirical world, and the method of scientific observation, however much maligned by 'runners' and 'stayers' alike, which alone provided the necessary knowledge for making the noetic attempt. He felt, perhaps more instinctively than rationally, that a teleological approach might rid him of this impasse. However much modern man may laugh, at the instance of Voltaire's *Candide*, about the idea that the empirical world should be the 'best of all possible worlds', Anaxagoras, who first proposed this, had a good case for maintaining that the noetic world was no more than the empirical world mirrored by the human logical faculties. That applied as much to the 'mathematical' world of the Pythagoreans as to the 'contemplative' world of Empedocles and the Eleatics, or even to the purely rationalist 'gnomic' ethics of Democritus. But it has to be remembered that Anaxagoras failed completely in his attempt to find any convincing argument why his active principle, the Nous, should, or even could, be immanent in the empirical world. For if there was nothing else than this empirical world, then there was no possibility for that reflection or comparison, which would justify his claim that it was the best possible world. This consideration, when brought up against Anaxagoras, even justified the concept of a moral dualism in the universe. Such a theory was indeed promoted at that time by his contemporary Empedocles. According to it an evil world had penetrated into a good one, and produced those limitations, which compelled even the optimist to restrict his verdict on the empirical world as 'the best possible'. It was indeed such ethical considerations which made it appear that there could be nothing good if there was

[1] Cf. H. Diels, *Der antike Pessimismus, Schule und Leben* I, 1921, and W. Nestle, *Griech. Weltanschauung*, 1946, 177 f.

nothing evil, which threatened to obliterate even the lasting physical recognitions at which Anaxagoras' optimistic teleological monism had arrived.

The man who went beyond Anaxagoras' profound demands, and somewhat sketchy proofs and conclusions, was, according to Plato, his erstwhile pupil, Socrates. We may not be able to reconstruct his cosmology; but the following three conclusions which Plato proposed under his name seem also to have been drawn from his doctrine. The first is that the dualism between the noetic and the empirical world was in fact logically inescapable; but that it did not follow from this recognition at all that it was the noetic world which mirrored the empirical world. Logically the opposite was to be preferred. This consideration then gave rise to Plato's doctrine of a cosmos of ideas the beginnings of which go back undoubtedly to the teaching of Socrates. The second is that this doctrine recommended itself particularly in view of the inadequate conception of the Nous held by Anaxagoras, and showed its virtue when compared with his inept conception of goodness. I believe that it was Socrates who insisted that the essence of the Nous was not its creative activity but its combining mathematical reason with non-mathematical contemplation. Plato concluded from this that the conceptions of the noetic world drawn by the Pythagoreans and the Eleatics were not mutually exclusive, and therefore to be rejected, but complementary and thus both necessary to the cosmos of ideas. Finally, the third conclusion was that the chief characteristic of the empirical world, shown by its constant motion and change, was its life, which was the source of this motion and change. Life, however, was not immediately attributable to the intellectual powers of the Nous, but to the soul. The soul was the 'beginning' of the empirical world in the sense that it established the connection between the two partners in the universe, the empirical and the noetic world, by making the life in the empirical world possible, causing motion and change as the self-moving mover. It was at this point, however, where Plato found an insuperable difficulty. Since the empirical world was continually functioning, in constant motion and change, no ontological predication from within it could be made about it. For the concept of existence includes of necessity an element of being static, unchangeable. Already Parmenides had had to face this difficulty; and Plato took it up in his *Parmenides*. The Eleatic proposition in this dialogue, developed by Zeno, had been that the motion and change in the empirical world were not real; yet, if our exegesis is correct, Plato's Parmenides himself admitted that Zeno's axiom ἓν τὸ πᾶν, the universe is one, was open to criticism. It is evident, however, that Plato did not at once find a solution to this *aporia.* The attempt which he made in the *Phaedrus*, granting to each individual human soul an unlimited transcendency, might have drowned all cosmological thought in utter subjectivism. In the *Timaeus* we find that the idea of order, τάξις, first proposed by Anaximander, was re-introduced by Plato in order to provide the static element in the ontology of the continually moving empirical world. It was the 'well-ordered' rotatory motion of the cosmic soul, imitated by that of the archontes, the rulers of the astral spheres, which caused the continuity of change and motion, and was thus the 'beginning' of the empirical world. However, whilst here a support for the insecure bridge between the noetic and the empirical world was found, it was at the same time threatened by the ethical dualism which we have found in Empedocles. Plato hinted in various mystical and dark allusions at the existence of a second 'beginning', a connection between that part of the supernatural world which is chaos, and that 'disordered motion' within the empirical world which is the source of all evil. We are given to understand that the connection is of an earlier age than the 'beginning' of well-ordered motion. Attempts have been made to connect this doctrine with Orphic mysticism and/or Iranian dualistic theology. However that may be, one fact is evident: the theodicy aimed at when Anaximander introduced the term of ἀρχή was brought into jeopardy by the *Timaeus.*

At any rate it has to be stated that Plato completed his doctrine of the 'beginning' in a cloud of mythology. He firmly maintained that the 'beginning' which he meant, was the beginning of motion, i.e. of the empirical world, the world of life; but he shrouded it with mysticism. For he was essentially what we would call a metaphysical thinker, and deeply conscious of the part played in life by metaphysical influences. It is a veritable puzzle when we see him, at the height of his life and mental power, give in so submissively to 'father Parmenides' as he appears to do in that dialogue. I think it probable that this submissiveness marked his recognition that the great concept of a cosmos of ideas could never be established on a basis of reasoning only. The problem of the *Parmenides* was not so much ontology as such, but its method. In a world of constant change and death, which nevertheless remained in existence, philosophy could not by-pass the necessity of defining what was meant by 'existence', not only of the individual but also of the entire universe which included, yet transcended, the empirical world. Plato realized that it had been the methodological question of the 'being of being', which had agitated the thought of the historical Parmenides; and it was this problem of how to make the existence of the universe plausible to the human mind, which he tried to answer in the myth of the *Timaeus*. The effect created by his answer, as soon as it was given, was immeasurably great. It took the sober realism of a genius like Aristotle to resist its magic, and to perceive not only that it was insufficient, but that it was unreal like the shadow of a dream.

It is evident that the split between Aristotle and the early Academy, embittered though it was by personalities, had its most profound cause in their disagreement over the *Timaeus*. Aristotle's praise for Anaxagoras, that he had been the only sober one entering the circle of the inebriated,[1] especially when it is contrasted with the immediately following reproach for the same philosopher's arid doctrine of the Nous,[2] should be seen in a contemporary setting. It signifies Aristotle's verdict that the myth of the *Timaeus*, and the speculations built upon it by Plato's successors, paid far too little attention to the methods of scientific observation, and particularly to the principle of causality introduced by Anaxagoras. And the lasting hatred which Aristotle felt for the Pythagoreans, has to be ascribed to the fact that their conviction that number was the subject in the universe, and all physical data no more than accidental attributes, opened the way for all sorts of superstition and magic, and almost barred it against the only admissible argument, the inductive argument of true science. Here too it was Speusippus and the early Academy which favoured Pythagoreanism, nay, it had been Plato himself who had accorded philosophical status to their esoteric delusions. Against such adversaries Aristotle established his system of instruction, to counter the development of Plato's doctrines which was going on in the *a priori* speculations of the Academicians. This system consisted mainly in a detailed scrutiny of the empirical world, and in the thorough examination of the workings of reason, as being the instrument given to man for this purpose. It is not surprising that in his reserve to the domination of *a priori* speculation going on in the Academy, Aristotle should subject the concept of the 'beginning' to such an acid test as he did in the eleventh book of the *Metaphysics*. It is even understandable, if illogical, that he tried, in the third book of the *Metaphysics*, to arrive at a precise conception of the 'beginning'—if such an idea was to be entertained at all—and not only at a lexicographical description of the word,

[1] Aristotle, *Metaph.* I.3, 984b, 15.

[2] Ibid., I.4, 985a, 18.

from the linguistic observation of the common use of the word. Apparently he did not take into account that anything observable would have to be of necessity dependent upon the 'beginning' for its very essence, so that it cannot contribute anything to its definition.

Remarks like these about Aristotle's erection of his philosophical system have to be judged in the light of two facts. The first is that the extant remains of his literary work allow us to observe the intellectual progress which he made until his thought reached its final maturity. This progress took the form of a gradual dissociation from Platonic thought, coupled with the very painful process of severing the ties with the Platonic companionship. These personal ties were of great strength. For he was married to the daughter of Hermeias of Atarneus who had also been Plato's pupil; and he had shared the years on the shores of Mysia in close friendship with Xenocrates, who was to become the head of the Academy as the successor of Speusippus. The second fact to consider is the strength of the conviction which forced him to break away not only from the Academy, but also from its teaching. When considering this point it has to be remembered that Aristotle's secession did not lead to an abrupt 'burn what thou hast worshipped, and worship what thou hast burnt'. To the end he shared with the Academy numerous, and indeed some basic, convictions. Perhaps the most important of these was the conviction that philosophical monism was untenable. It was also of great significance that he maintained to the end that it was the cosmic soul which was the origin of the life of the empirical world. But even the differences in the cosmological doctrine were to a large extent differences in delimitation. There was the question whether or not the empirical world coincided with the physical world, so that the moral world of man had to be regarded as a separate entity. For the Platonic cosmology the question was unanswerable. For whilst 'that which is just by nature' played its part in his *Republic*, so that the autonomy of the realm of morals had to be presupposed, the well-known fact that Plato nowhere made use of the words *φυσικός* and *ἠθικός*,[1] warns us from drawing far-reaching conclusions from his conception of 'natural justice'. Aristotle, on the other hand, went at least very near to a complete separation of the physical from the ethical realms. Again there is the problem of the distinction between the supra-lunar and the sub-lunar world. There seems to be no definite distinction between the two in Platonic cosmology; and his Socrates even refuses to recognize either

[1] Cf. Liddell–Scott, ss.vv.

the sun or the moon as gods.[1] Nevertheless, the increasing importance for Plato of astrological considerations, which finally made him establish 'the meaning of things existing' in the stars,[2] makes us wonder in what sense he may have held that the celestial phenomena should be regarded as belonging to the empirical world. Aristotle, on the other hand, who established the distinction, can be said with some assurance to have held that, because the stars are visible, the trans-lunar world should be regarded as part of the empirical world, although there are occasional astrological remarks occurring in his *Metaphysics*.

These differences may appear small, and even paltry, when they are taken by themselves, but they may serve as straws in the wind. For it was in the field of cosmology where Aristotle parted company with the early Academy; and the overruling conviction which made him do so, was that the 'beginning' had to be clearly understood as causation. It is now generally accepted that the early Academy was right to take the creation passage in the *Timaeus* allegorically. Why then was it taken literally by Aristotle? Does it really seem probable that he, the personal pupil of Plato when he developed his cosmological thought in his lectures, did so erroneously? I think not! He found there a weapon for the defence of his causal cosmology, and he used it. But even if it should be true that he made a mistake, it was a revealing mistake. For it laid bare the reserve shown almost instinctively by Plato and the early Academy to the causal understanding of the 'beginning'. Aristotle felt no such misgivings. In fact, his proposition of a passive causality of 'matter', which in the earlier systems of Greek philosophical cosmology, and particularly in Plato, is not forthcoming, seems to transcend by far the limits of inherited, unphilosophical theories of causality.[3] He was also quite clear about the effect of the co-operation of the active and the passive causes, which was motion and change. And finally he maintained very strongly that every causal nexus must show itself in a limited series

[1] Plato, *Apol.*, 26D, where the material character of sun and moon is emphasized.

[2] Plato, *Legg.* XII, 967D.

[3] The best instance of an unphilosophical theory of causality in a cognate system of thought comes from Roman Law, where the *lex Aquillia* made the defendant liable for the *damnum corpore corpori datum*. No co-operative passive causality contained in Aristotle's 'form' and 'plan' is looked for. The ancestry given for the 'passive cause' by F. Solmsen, *Aristotle's system of the physical world*, 1960, 118 f., presupposes for most of the Pre-Socratics and Plato a system of causal thought. Even after Anaxagoras that is a *petitio principii*, and how much the 'passive causality' is Aristotle's own property appears clearly from his polemic against Plato, *De gen.* I.2, 315b, 25 f.

of events, so that the scheme of 'beginning, middle, and ending', which had been rejected with so much energy by Plato in his *Parmenides*, was re-instituted in Aristotle's cosmology. For it was logically proved by him that an infinite series of events could not be meaningfully explained in terms of cause and effect.

In all these considerations Aristotle confined himself deliberately to the empirical world, so much so that, as we have seen before, he even tried to derive the meaning of the 'beginning' from the use that was made of the term in common speech. His attack upon the earlier philosophers in the first chapters of his *On-coming-to-be* illustrates well his objections to any philosophizing that proceeds from a basis of insufficient knowledge. However, he was not forgetful of the part which the noetic element had to play in his as in any other discussion about the 'beginning'. He rejected decisively the Platonic doctrine of a world of ideas, which would exist as the cause pre-forming the empirical world. Instead he established the claim that the empirical world was based upon truth. This claim was derived from his definition of subjective truth. By it he maintained that statement arrived at by the way of observation of a critically defined phenomenon of the empirical world, presupposing the law of causality, and with the help of logical deduction connected with previously established knowledge about kindred phenomena, is the truth about that part of the empirical world with which it deals. Although Aristotle may not have said so with so many words, this definition of subjective truth meant that an observation laid down in a true statement by a person A could be made by a person B also. True observations of nature were repeatable. This in its turn proved that the empirical world was organized upon a principle of objective truth, which corresponded with the subjective truth of the critical observer. And that meant that the empirical world was in reality a world of causality and truth. This conception of truth, therefore, introduced a static element into the empirical world, that static element for which all the earlier philosophers had been clamouring. However, although truth had this function in the empirical world, which truly existed in the same way in which a true statement could be made about it, it was not of the empirical world, for it could not be apprehended by sensual perception. Wedded to the doctrine of causality, it had to be derived from an ultimate source, just as causality from an ultimate cause. And there could obviously be no more than one ultimate. However, truth and causality are an unequal match. For truth, as we have seen, was the static element, was the same always since its chief characteristic was that it was repeatable. Causality, however, had its essence in motion. And yet they had to refer to the same supreme authority. The logical regress which we have pointed out demanded one supreme solution by which those two could be combined. This solution was found by Aristotle in the figure of the unmoved mover. He had to be unmoved because he was infinite truth; and he had to be the mover because the very existence of the incessantly moving world was brought about by him. However, this process of a continuous creation was by no means uniform. Life, in particular, was mediated to the cosmos through the self-moving mover, the cosmic soul.

It can hardly be imagined that this system has even yet lost its impressiveness. However, it cannot be denied that European philosophers have for centuries worked in order to show the many questions it left unsolved and the numerous gaps which it contained. These gaps and questions are not only concerned with details but also fundamental ones. Arisotle's system is often described as an ethical cosmology; and therefore something will have to be said about Aristotle's conception of the moral cosmos, which he found in the world of man, thus separating it from the empirical cosmos as a whole, and relegating the conceptions of good and evil to the world of man. Here, if we are not mistaken, lay yet another deviation by Aristotle from the teaching of the Academy. Plato, we remember, had still connected evil with his conception of disordered motion; and the cosmic origin of evil had played its very important part in the cosmology of Empedocles, upon which Plato may have depended in this matter. Aristotle, on the other hand, became the father of the later Stoic conviction that the good is that which is congruent with human nature, that man is good by nature. With this addition, we may state that after a most intense struggle Aristotle succeeded with his intention of taking into account the entire discussion of the earlier philosophers on the question of the 'beginning'. If it is held, as it should be, that he unduly stressed his thesis that the 'beginning' was the supreme cause, it should not be forgotten that he allied to this conception that other of the supreme truth, which was meant to embody the earlier and perhaps more fundamental meaning of 'beginning', the meaning of 'principle'. It was only the slackening of his successors, Peripatetics and Stoics alike, which brought this consideration into oblivion; and subsequent generations of European philosophers have repeated it over and over again that the first 'beginning' is the supreme cause, is god.—And the theologians have applauded them.

9 Hellenistic developments of the doctrine of the 'beginning', and the *creatio ex nihilo*

I

With the cataclysm of classical Greece the type of philosophy of which Aristotle had been the last representative disappeared. Thinking for its own sake, without a material return, even though it showed its practical value occasionally by predicting an eclipse of the moon or the sun, was replaced by a dogmatism which had to answer the question: What good is this going to do? It has been held that not only the Aristotelian philosophy, but indeed every branch of Greek philosophy since Socrates, took its cue from an ideal of eudaemonism, general happiness, and in this sense had its centre in moral philosophy. Probably this was true of the Cynic and Cyrenaic schools; but to my mind neither Plato nor Aristotle can be justly charged with such a betrayal of true philosophy. They concentrated their thought upon man, knowing that in order to find the truth about something it is not only the subject of enquiry that has to be examined, but that the enquirer himself as well as the act of observation has to be brought under the light of critical examination. Their purpose, however, was the establishment of objective realities, the deity and truth. It was the period after the conquest of Persia by Alexander's irresistible forces which changed the much maligned axiom of Protagoras that 'man is the measure of all things' to the truly questionable programme that man's welfare should be the measure of all things. In this context the problem of the 'beginning' as such ceased to be of any special interest; but it became important as the problem of the beginning of something in the sense of the popular Greek proverb, known already to Plato, that 'the beginning is half of the whole'.[1] Its importance was thus derived from the fact that an answer might provide an insurance against the unexpected.

There was only one neutral field left where the discussion of the 'beginning' was allowed to continue, as far as can be seen, on the same lines as before, the field of mathematics. Here we find Aristoxenus of

[1] Leutsch–Schneidewin, *Corpus paroem.* I, 213, Diogenianus II.97, with numerous parallels.

Tarentum, a pupil of Aristotle but with strong leanings towards Pythagoreanism, proclaiming that 'the monad is the beginning of number';[1] and there is also a fragment of Neo-Pythagorean origin, ascribed to Boutheros, one of the master's own pupils, 'the One is substance and nature and mind and instruction, for it holds the beginning, the middle, and the ending'.[2] Such sayings still held a significance which went beyond the realm of technical mathematics. Unfortunately they were as a rule rather heavily scented with magic,[3] and in that way rather threatened than protected their adepts.

Two ways in particular were open in Hellenistic philosophy to lead to the welfare (*εὐδαιμονία*) of those who took them, the one leading to collective, the other to individual welfare. The latter had been indicated already by the Cynics and Cyrenaics of the fourth century B.C., but was now more elaborately outlined by Epicurus. The former, taking its matter from all the former Greek schools of philosophy, with an admixture of oriental mysticism, found its expression in Stoicism. Stoicism was therefore a utility philosophy, as well as a utilitarian one. Cynicism continued also, developing more and more into the philosophy of the declassed; and Epicureanism—with the puzzling interlude of the Roman nobles, especially Cassius and others amongst Caesar's murderers, adopting it—became the philosophy of the well-to-do quietists and moral reformers. Stoicism had to provide a tolerable ideological background for the despotism of the Hellenistic monarchies; and the longer these continued in their internecine wars the louder became its demand for peace, until the Roman principate took this demand literally. For the philosophical peace of the Stoics was to be peace in the universe, in nature. Nature was by no means regarded as obviously peaceful. Especially the mystics would represent it as a hostile, irrational, and irresistible force, threatening human well-being.[4] The Stoic reply to such

[1] Aristoxenus in Stob., *Ecl. de arithm.* 6, ed. Wachsmuth, I, 1887, 19. On Aristoxenus cf. Überweg–Prächter, *Philos. d. Altertums*, 12th ed., 1926, 404.

[2] Boutheros in Stob., *Ecl. de arithm.* 6, ed. Wachsmuth, I, 19. Of Boutheros only the name is known from the list of Pythagorean apostles in Jambl., *Vita Pyth.* 36.267, and his home town of Cyzicus. His saying belongs together with several other 'beginning-middle-and-ending' statements, e.g. Plutarch, *De facie* 11, 925 f., where it is denied that either the universe or even the cosmos have a beginning, middle or ending. P. Raingeard, *Le 'peri tou prosopou'*, 1935, 83 f., treats this as an Academic tilt at Stoicism, but the cosmological significance is much larger.

[3] Cf. Bidez–Cumont, *Les mages hellénisés* I, 1938, 42, *artes magicas invenisse et mundi principia . . . diligentissime spectasse*, from Justin, *Hist. Phil.* I.1.9.

[4] Cf. *Politische Metaphysik* I, 1959, 153 f.

assertions was the distinction between the universe and the cosmos. Human life was lived in the cosmos, and here there were two guarantees[1] for mankind. The creator, so Zeno proclaimed, and no Stoic teacher, down to Epictetus,[2] ever failed to repeat it, was the father;[3] and all that happened in the cosmos was effected by an active cause, which was god, and a passive one which was matter, and was altogether innocuous because it was incapable of any action of its own.[4] This god was further explained as the logos,[5] apparently in so far as he was immanent in the cosmos, and as Zeus. It has to be noticed, however, that according to the Stoics the first creative act, the separation of matter into four elements, was not taken by the logos.[6] The elements were again different from the *ἀρχαί* but in this respect a closer examination of Stoic physics is still required. What can be seen is that the 'creative fire', the physis of the Stoics, causes in the original matter an extreme heat, which in its turn is the *ἀρχή* for the element fire.[7] This 'creative fire', which is material, but divine, embodies all the 'germinating logoi' from which all things take their being according to fate or predestination (*εἱμαρμένη*).[8] The two kinds of fire, it seems, were passed lightly by Zeno's critics; but it was in his rigorous determinism that he and his school failed both the demands of logical thought and those of his followers.[9]

Together with the general rocking of the foundations of Stoicism the

[1] M. Pohlenz, *Die Stoa* I, 1948, 65, has described the Stoic scheme quite correctly; but he has not seen that it was only in the cosmos that the physis prepared the way for the logos, cf. e.g. Galenus, *Hist. philos.* 20, Diels, *Doxogr.*, 611.

[2] Cf. e.g. Epictet–Arrian I.3.1; III.24.16.

[3] Diog. Laert. VII.147.

[4] Cf. H. v. Arnim, *SVF.* I, 24, 5 f. (Zeno); ibid., 110, 25 f. (Cleanthes).

[5] Diog. Laert. VII.134, one of the fundamental doctrines of Zeno.

[6] This follows from a comparison of the quotations from Aëtius and Achilles Tatius in H. v. Arnim, *SVF.* I, 24, 9 f., which mentions the separation of the elements, but not the logos as effecting it, and Diog. Laert. VII.134, mentioning the logos as the active cause in the cosmos, but not the separation of the elements as its first act.

[7] Diog. Laert, VII.134 = H. v. Arnim, *S.V.F.* II, 111, 4 f., with a misleading reference in the index IV, 28a, 37, and Galenus, *De elementis* I.6, *S.V.F.* II, 134, 28 f.

[8] Material deity, Galenus, *Hist. philos.* 16, Diels, *Doxogr.*, 608, 16 f. Predestination, Aëtius, *Plac.* I.7.33, Diels, *Doxogr.*, 305 f.

[9] Cf. how Aëtius, *Plac.* I.28.3, Diels, *Doxogr.*, 323 f., notes the discordances in Chrysippus alone; neither does the gradation Zeus, nature, fate, ascribed by him to Posidonius, ibid. 5, Diels, 324, inspire greater confidence. However, the influence of this doctrine was still felt in a late period when the scholiast to Homer, *Il.* XVII.409, ed. E. Maass, VI, 1888, 229, recorded that 'certain philosophers called fate (*μοῖρα*) the logos of Zeus'.

Stoic doctrine of the 'beginning' came to grief because Zeno and his followers had put the causal conception of 'beginning', which Anaxagoras had introduced, in a purely materialistic form, cruder even than Anaxagoras had formulated it. They disregarded the thorough revision it had received by Aristotle, and which admittedly put the monistic materialism of Anaxagoras, already inconclusive in its original form, into jeopardy. The Stoics, for all their frantic assertions to the contrary, were forced by logic to face the following dilemma: either they had to make allowance for a cause outside their world of space, or else 'space' and 'substance' would have to part company, and the vacant space, which they regarded as the non-existent, would have to take the part of the active cause. Already Zeno attempted both solutions. The fire of 'nature', as we have seen, was not the element of fire, but this fact was only admitted for the first act of creation, after which the logos took over; but on account of this the Stoic logos, as we shall see, was derided as a true *deus ex machina.* Also the Stoic cosmos was surrounded by a universe, which in its turn led into vacancy; but thus their causality, which only worked by contact, was proved wanting. For how could a deity that embodied all the germinating logoi, the ultimate cause as Zeno described it, be represented at the same time as a vacant space, especially if this space was meant to be infinite? No contact cause was possible in the vacant space; and in the infinite no causality was possible at all, as Aristotle had proved. This dilemma was also the reason why the conception of matter became ambiguous, already in Zeno's own teaching. Quite apart from the divine plasma of the logos, he had to distinguish between substance, which he called the first matter, so passive that it could not even be perceived with the senses, and matter only, the stuff from which the empirical cosmos was formed.[1]

The Stoic cosmology was not improved either by their shibboleth, the doctrine of 'the great year', ending with the periodic conflagration of the cosmos and its subsequent restoration. Arius Didymus, the last systematician of the Stoa, said:[2]

For there has to be an underlying substance capable of enduring all changes, and the creative power working on it. For just as in us there is a creative nature, so there is something analogous also in the uncreated world. For the beginning (ἀρχή) of creation has no power over this kind of nature; and as it is uncreated

---

[1] Cf. J. Moreau, *L'âme du monde*, 1939, 159.

[2] Arius Didymus frg. 37, Diels, *Doxogr.*, 469 = Euseb., *Praep. Ev.* xv.19.3, ed. Mras, II, 383 f.

it cannot be destroyed, whether it be by its own act or by an external force aiming at its destruction.

There remained therefore, those two indestructibles, and from them started immediately a new world: 'For it is neither possible to bring the cause to a beginning or a rest, nor him who controls these things.'[1] As it stands this is, of course, nonsense, for logically there can be no cause without a beginning; but it is significant that this Stoic, at the time of Augustus, put, over cause and substance, a personal controlling deity, and considered the possibility of an external power interfering with them. There remained yet a popular Stoicism, in which any uncomfortable questions were answered with the blessed word 'nature'. Nature, we have heard from Arius Didymus, was exempt from all 'beginning' or coming to be; and this idea was probably a relic of Panaetius' complete rejection of the theory of periodical conflagration.[2] Posidonius, restoring it, stated such an exemption for three great cosmic powers, Zeus, nature, and fate (*εἱμαρμένη*);[3] but it seems clear from Didymus that the pre-eminent term was nature. 'What else is nature', says Seneca, 'than the godhead and the divine reason, immanent in the whole world and its parts.'[4] What an edifying oratory! but when it came to hard facts the great moralist had to admit: 'It is so much easier to understand nature than to put it in words.'[5] The Stoics of today are still trying to put it in words why their 'nature' should not be *ἄλογος*, meaningless, like that of the Epicureans.[6] In the meantime not only their case went by default, but with it that of the Greek view of nature. For this, says R. G. Collingwood, 'had been based upon the principle that the world of nature is saturated or permeated by mind'.[7] Such indeed had been Zeno's conception of the cosmos, which by the beginning of the second century A.D. had finally lost not only its philosophical reputation, but—even worse—its interest. Stoicism, perhaps already from the days of Posidonius, 'was no more than a religion, in which every tendency of

[1] Ibid. § 2.

[2] Cf. the frgg. 64–66 and 68/9 in M. van Straaten, *Panaetii Rhodii fragmenta*, 1952, 19 f. The mention made of Panaetius' name in frg. 67, Arnobius, *Adv. nat.* II.9, as one of the upholders of the doctrine, only goes to show the general muddled-headedness of this 'Christian' father.

[3] Aëtius, *Plac.* I.28.5, Diels, *Doxogr.*, 324, cf. Galenus, *Hist. philos.* 42, Diels, ibid., 620, 20 f.

[4] Seneca, *De benef.* IV.7.1, cf. *Quaest. nat.* II.45.3.

[5] Seneca, *Epist.* CXXI.11.

[6] Cf. Aëtius, *Plac.* II.3.2, Diels, *Doxogr.*, 330b, 3.

[7] R. G. Collingwood, *The idea of nature*, 1945, 3.

accommodating the world to humanity had been extinguished, and all that mattered was to provide the guidance for the most patient endurance of things as they are'.[1] Even if it was not forgotten in the post-Christian era that once upon a time Zeus had been described not only as 'the mind of the cosmos' but even as 'the soul of the cosmos in form of air', there can be little doubt that the philosophical meaning of these assertions were lost.[2]

II

If the Stoic doctrine of the 'beginning'—and even of the 'beginnings'[3] had been a simplification from the start, and a simplification of the most serious kind, it has to be acknowledged that Stoicism recognized its obligation to provide the specialist sciences with a philosophical foundation, and tried sincerely to honour it. Consequently it was a comparatively large number of scientists, Eratosthenes the geographer, Euclid the mathematician, Celsus the physician, to mention only a few,[4] who professed their adherence to Stoicism. The evidence from the doxographers for the manner in which the fundamentals of physical science were discussed in other circles during the Hellenistic period is indeed disheartening, and may provide a fitting commentary to the well-known complaints of the elder Pliny about the decline of the natural sciences in his time. Already of Theophrastus it can be held with certainty that he introduced the word ἀρχή in his summaries of the doctrines of the natural philosophers, although it was not to be found in the original. That was done knowingly in the case of Thales, who had tried to reduce all the manifold physical phenomena to one common denominator, water. Theophrastus called it ἀρχή, although he admitted that the term had only been introduced by Thales' pupil, Anaximander; and we have found good reason to believe that the natural philosophers after Anaximander avoided the term again. Such lack of precision was justified by Hellenistic philosophers with Socrates' saying that the enquiry into the ultimate

[1] H. & M. Simon, *Die alte Stoa und ihr Naturbegriff*, 1956, 95.

[2] Zeus as the nous, Homer, *Il.* XIV.252 schol., as the soul, ibid. XVI.233 schol., ed. E. Maass, VI, 1888, 79, 176. On the related Stoic doctrine of the *pneuma* cf. J. Moreau, *L'âme du monde*, 1939, 162 f.

[3] Cf. on the 'beginnings' Galenus, *De elementis* I.6, *supra*, p. 156, n.7, and F. Lammert, *Hermes* LXXXI, 1953, 488 f., with references.

[4] Cf. P. Barth, *Die Stoa*, 3rd and 4th ed., 1922, 147 f. (The later editions of this book, by A. Goedeckemeyer, cannot be recommended.)

reasons and parts of the things was beyond human powers, and of no practical value.[1]

It seems that the particular interest which caused the frequent use of the term was political. It was the monarchy, *μία ἀρχή*, of the cosmos which those Pre-Socratics had asserted or rejected.[2] Theophrastus' other interest, and here again his Hellenistic attitude was clearly shown, was the history of the Greek philosophical schools, and in particular the succession of their heads. The whole organization of his *Physikōn doxai* was personal rather than material. The revision of his work, of which the fragments from Aëtius (*ca.* A.D. 100) give an idea, concentrated even more upon this subject, and the approach was often critical.[3] These later Hellenists tried to assert themselves, and their criticism even turned into invective when it came to a figure like Plato.[4] Later on, and presumably only in excerpts from Aëtius' work, a kind of tabulation was used. The form, 'Pythagoras says that the world was created rationally, not temporally; Plato says . . .; Heraclitus says etc.'[5] made its appearance. These may have been meant as text-books for schools; but they were not much below the real disputations, which served as a pattern for Seneca's, and even Cicero's, dialogues. The Roman *jeunesse dorée* cracked their jokes about them: Propertius said that when he would be too old to love, he would study the theory of nature.[6] However, the elderly among the Roman nobles did enjoy in their *villeggiatura* a discussion on this very subject.[7] The one or other of them may even have been committed to this or that Greek

[1] Cf. L. Friedländer, *Sittengeschichte Roms III*, 9th ed., 1920, 275.

[2] Cf. esp. Theophrastus frg. 1, Diels, *Doxogr.*, 415, 1 f. A similar interest is underlying frgg. 2 and 5, Diels, ibid., 476, 3 f.; 480, 4 f., whilst frgg. 3 and 4 deal with polyarchy.

[3] Characteristic are remarks on Anaximander and Anaximenes, Aëtius, *Plac.* I.3.3, 4, Diels, *Doxogr.*, 277, 15 f.; 278a, 16 f.; b, 12 f. They were suppressed, I believe, by Philo and Ps. Justin, *De monarch.*, Diels, ibid., 278n., not added by Plutarch.

[4] Aëtius, *Plac.* I.7.4, Diels, *Doxogr.*, 299, 10 f., cf. ibid. 7, Diels, ibid., 300, 4 f.

[5] Aëtius, *Plac.* II.4.1 f., Diels, *Doxogr.*, 330 f.

[6] Propertius IV.4.23 f.:

Atque ubi iam Venerem gravis interceperit aetas
Sparserit et nigras alba senecta comas,
Tum mihi naturae libeat perdiscere mores,
Quis deus hanc mundi temperet arte domum etc.

The joke lies, of course, in the double meaning of *natura* as 'sex' and 'empirical world'.

[7] Cf. Cicero, *De nat. deor.* I.2, 19, 53 f.; II.97.

philosophical school, Stoicism or even Epicureanism, but it cannot be assumed that such a company would have launched out into original thought. At best they would have with them their domestic, mainly Greek, philosophers,[1] from whom they received their instruction, presumably with the help of doxographical hand-books, about the problems of natural philosophy, their originators, and how to argue about them. No special knowledge was required. The pious Vergil, for instance, is said by his commentator Servius to have used repeatedly in his poems the cosmology of the atheistic Epicureans.[2] Was he then guilty of hypocrisy, or should we not rather conclude that he objected to drawing logical consequences as much as any well-bred Englishman? It cannot be denied that this whole literature of letters, dialogues, declamations, consolations, anthologies and all the rest, gives the impression of being desperately insincere. It is, therefore, not easy, even when confronted with the most learned attempt at reconstructing the philosophy of a master like Posidonius from such material, to subdue an ultimate feeling of no confidence.

There was yet one particular field of science where insincerity was inadmissible. Serious study was demanded and, what is more, paid handsomely in the medical faculty. I venture to doubt whether the study of medicine nowadays opens the way to a philosophical cosmology. It is true that schools of medicine demand of their students a minimum of knowledge in general sciences. However, these subjects are easily the least popular with the present colleagues of Hippocrates, and unlikely to make them take an interest in the philosophy of healing, let alone of the cosmos. However, ancient Greek cosmology from its very beginning had stated as an axiom that there was an analogy between the cosmos and the human body; and the cosmology of the Stoics in particular insisted upon the correctness of this claim with such energy that it has been described with good reason as a 'cosmo-biology'.[3] In view of the fact that the 'beginning' and the evolution of the human body in the maternal womb, although it was a mysterious process, was a common occurrence, the question whether the cosmos had been created was of immediate reality in these medical circles, and was

[1] Cf. how Scipio, Cicero, *De rep.* I.15, complains about the absence of 'our Panaetius'.

[2] Cf. Servius ad Vergil, *Ecl.* VI.30 f.; ad *Georg.* II.336, ed. Thilo, III, 1887, 69 f., 247 f.

[3] J. Moreau, *L'âme du monde*, 1939, 166 f. Cf. ibid., 131 f., where the author outlines how much not only Aristotle but already his authorities had done in this respect for the preparation of the Stoic doctrine.

discussed there with the thoroughness which their science demanded. They also tried to examine what the cosmos might be created of, and what was its nourishment. This was the reason why that great physician Galenus showed such a lively and, on the whole, well-instructed interest in the theories on the creation of the cosmos.[1] The bulk of his information was presumably taken from Theophrastus, of whose exposition one fairly substantial fragment is still available.[2] It is, however, significant that in the revision of Theophrastus' encyclopaedia by Aëtius there seems to have been no special chapter dedicated to the question whether or not the cosmos had been created. Of course, the question is discussed time and again, but not in a special chapter; yet there are traces that Theophrastus' work had contained such a chapter.[3] However, in the first century A.D. this question was of an immediate interest only to the physicians and—the Jews;[4] and apparently Aëtius was neither.

III

The problem of the creation of the world did yet touch the educated people in general during the Hellenistic and Roman period when it was put in a special way: What about the *creatio ex nihilo*, creation from nothing? Galenus is here full of valuable information. For him the question offered an opportunity for a gallant attack upon those philosophers who believed in a deity specially produced for the purpose of acting as the creator.[5] Contemporary mystic doctrines abounded with such deities; but I prefer to believe that the adversary thus floored was the Stoic doctrine of the creative activity of the logos after the conflagration. It is true that Galenus neither here nor elsewhere says 'from nothing'. This idea, if it is an idea, has to be supplied from Lucretius, *De rerum natura*, especially the line I.150, *nullam rem e nilo gigni divinitus umquam*.[6]

[1] Galenus, *Hist. philos.* 17, Diels, *Doxogr.*, 603.

[2] Theophrastus, *Nat. opin.* 12, Diels, *Doxogr.*, 486 f.

[3] Homer, *Od.* I.263 schol., ed. Dindorf, says 'there are views expressed about the cosmos. One of them holds that the cosmos has been created and will perish etc.'. That is a typical opening of a chapter, and where else than from Theophrastus in the last instance could the scholiast have obtained it?

[4] Cf. esp. the ps. Philonic *De aeternitate mundi*, in which the fragment from Theophrastus, *supra*, n.2, has been preserved.

[5] Galenus, *Hist. philos.* 17, Diels, *Doxogr.*, 609.

[6] Actually the five lines I.146–150, ought to be quoted. Cf. J. Bernays, *Über die unter Philos Werken stehende Schrift von der Unzerstörbarkeit des Weltalls*, Abh. Akademie Berlin, 1882/3, 12 f., on the importance of *divinitus*.

It must be said, however, that absolute nothing is hard to understand; and if Galenus and his allies did not assume that it was 'from nothing' when they wrote 'creation', the acerbity of their polemics would be hard to understand also. It has to be remembered that from the time of Plato and Aristotle the Greeks knew that matter as such, being absolutely unobservable, does not exist. Thus we hear,[1]

> and again that god whom they (Anaxagoras and Plato) call, either did not exist himself when things did not move or were moving in a disorderly fashion, or he slept, or was just waking up, or doing neither,

or a Neo-Pythagorean forged a fragment in the name of Aristaeus, supposedly the immediate successor of the master,[2] in the following terms:[3]

> If a beginning of the universe is assumed, then that which moves it must have fallen asleep at a pause of the universe for rest. If then the mover could sleep or rest he would be perishable and created, and would himself have a limit to his motion as much as the whole, the entire universe. This calls for a decision between no creation or a creation out of nothing.

We can hardly assume that these authors had any intention of imitating the prophet Elijah (I Kings 18:27); but we have seen how Arius Didymus, the great Stoic authority, defended the doctrine of periodical conflagration to give the cosmos rest, asserting violently that the Stoic deity did not rest—even if the logos vanished with the cosmos. Here indeed lay the deepest root of the conflict. From here we can measure the depth of defeat into which Stoicism was plunged when, in the middle of the second century B.C., it had to appoint Panaetius as head of the school, the man who had publicly disavowed the doctrine. From here we understand why Posidonius, when re-establishing this ancient Stoic dogma, had to devote considerable space and energy to discussing creation from nothing, and why he used a typically Epicurean expression when he declared it as 'unreal'.[4] However, it seems doubtful whether this defence availed much against an adversary who stuck to his thesis that 'nothing comes from nothing. For all comes from all,

[1] Aëtius, *Plac.* I.7.8–9, Diels, *Doxogr.*, 300.

[2] Cf. Jamblich., *Vita Pyth.* 265, see also ibid. 104.

[3] Aristaeus in Stob., *Ecl.* I.20.6, ed. Wachsmuth, I, 1884, 177.

[4] Posidonius in Stob., *Ecl.* I.20.7, ed. Wachsmuth, I, 177 f. The fact that the word *ἀνύπαρκτος* used there is Epicurean has been stated in Liddell–Scott s.v., on the evidence of Epicurus frg. 27, ed. Usener, *Epicurea*, 100 = C. Bailey, *Epicurus*, 120, frg. B 3.

and no seeds are needed'.[1] For here the great Stoic doctrine of the *logos spermatikos* was rejected once and for all.

Once more it is necessary to remember that terms like 'non-existent' or 'nothing' are equivocal.[2] It is important for us to see what lay behind the words 'creation from nothing', which were so hotly discussed in later Hellenism. The great fear of the time was the fear of the last day which would bring doom to all mankind, and thus reveal the utter senselessness of all human life. It is childish to monopolize this threat for Jews and Christians, but they claimed to possess particularly effective antidotes, a salvation which would preserve the soul of the individual unto eternal life. The fear of the coming destruction of the universe was not, however, tied to any one religion nor even caused by religion. To understand it properly we have to remember how small, comparatively speaking, the *oecumene* was at the time of Cicero. It was an island surrounded by the fury of oceans, deserts, and barbarians, all equally lawless; and the number of people keeping the *oecumene* civilized was shrinking. Under such circumstances not only the idea of doomsday was terrifying, but equally so that of the primeval chaos. But this idea was necessary to consolidate the comforting doctrine of progress.[3] The kindly physician Galenus accordingly sugared the pill scrupulously, before administering it. His description, taken as he claims from Plato,[4] is thus a mild one:[5]

There was nothing earlier from which any 'beginning' could have come; but it so happened that prior to the elements there was some invisible, shapeless substance, which the ones call qualityless matter, and the others actuality and deprivation (ἐντελέχεια and στέρησις). Now the elements are composites and products (ἀποτελέσματα); but the 'beginning' is nothing of the kind.

The description of the situation 'before' the beginning—all these thinkers are wedded to thinking in terms of temporality—could be much grimmer. There is another one which still tries to avoid mythology in the

[1] Epicurus, *Epist.* I.38 = Diog. Laert. x.38, ed. Usener, Epicurea, 5, 13 f.; C. Bailey, *Epicurus*, 20, 10 f. The impact made by Posidonius upon his Epicurean adversaries, which was not negligible, may be seen from their increasingly numerous professions of piety after the assault, cf. H. Diels, *Doxogr.*, 127 n. 1.

[2] A. Ehrhardt, 'Creatio ex nihilo,' *Studia Theologica* IV, Lund, 1950, 31 f.

[3] Cf. the profound philosophical analysis of this typically Hellenistic problem by Ernst Hoffmann, *Platonismus und Mystik*, Sitzungsber. Akad., Heidelberg, 1934/5, II, 82 f.

[4] The reference is obviously to Plato, *Tim.*, 30A.

[5] Galenus, *Hist. philos.* 21, Diels, *Doxogr.*, 612 = Plutarch, *De plac. philos.* I.2.

Homeric allegories by Heraclitus, possibly a pseudepigraphon of the time of Augustus:[1]

For there were old times when there was only shapeless matter, not yet coined into any final form. Not yet had the centre of the universe been firmly fixed upon the earth as its seat, nor did the heaven turn, establishing a road round it. All was sunless, vacant, and dim silence; and there was nothing more than the formless impassivity of outpoured matter, before that creative 'beginning', the world-mother (κοσμότοκος), drawing out the salutary model for life gave order to the cosmos.

This saying tries to be philosophical, but religious sources are just round the corner. Parallels to Gen. 1:2 are evident, although an immediate influence seems out of the question; and the 'outpoured matter' brings to mind an Orphic doctrine which has been preserved in a Judaeo-Christian document, the sixth pseudo-Clementine homily:[2]

Since the fourfold-begotten (τετραγενής) matter is animated (? inanimate), and a whole bottomless pit is ever running and streaming without discretion, and spouting up ten thousands of indeterminate mixtures, therefore dissolving itself in disorder and gaping as if unable to bring forth a living being, it almost accidentally conceived from the spirit brooding over it the cosmic egg.

The text seems to be in some disorder; but the passage is most valuable, and not only as a parallel to Gen. 1:2. It was meant as an exposition of ancient Greek religious wisdom, by the pagan philosopher Appian; it quoted a verse from Homer, *Iliad* VII.99, to which there exists a scholion which supports what is said in the homily with a quotation from Xenophanes, the pre-Socratic philosopher and founder of the Eleatic school,[3] and there is even the—not altogether trustworthy—allegation by Epiphanius that Epicurus had adopted the doctrine of the cosmic egg.[4]

We have been led, almost unawares, into the realm of mythology, the boundaries of which are very fluid where the creation of the world is concerned. We should rather consider the question whether there is

[1] Heraclitus, *Homer. alleg.*, *De Proteo*, ed. Th. Gale, 1688, 489 f.

[2] Ps. Clement, *Hom.* VI.4.1, ed. Rehm, I, 1953, 107, 10 f. = O. Kern, *Orphic. Fragm.*, 1922, 132, frg. 55.

[3] Homer, *Il.* VII.99 schol., ed. E. Maass, V, 1887, 239, cf. Diels–Kranz, 5th ed., I, 135 f., Xenophanes frgg. 27 and 33. Ps. Clem., *Hom.* VI.3.2, ed. Rehm, 107, 2 f.

[4] Epiphan., *Adv. haer.* I.8, Diels, *Doxogr.*, 589, 11 f. I cannot find that either H. Usener, *Epicurea*, 1887, or C. Bailey, *Epicurus*, 1926, has taken notice of this rather staggering assertion. Bailey does not mention Diels, *Doxogr.*, in his bibliography; but Diels has made copious use of it. I conclude that he found the report not even worth mentioning, and in this respect I disagree.

evidence for the existence of a Greek philosophical doctrine of creation from nothing. In an essay which was published several years ago,[1] I have drawn attention to two Peripatetic and one Platonic source which furnish the evidence for an affirmative answer to this question. I can now add another Platonic source to the list, coming from one of the last representatives of the 'middle Academy', Atticus, a contemporary of Galenus in the second half of the second century A.D. He says that 'the cosmos was made the noblest work by the noblest workman, who granted power to the creator of the universe by which he made the cosmos which previously was not'.[2] All these sayings have the purpose of claiming that the creation out of the non-existent brought forth reality; and this is the true problem of *creatio ex nihilo*. Once this problem has been faced, it is easy to see whence it took its origin. It was the Eleatic school which, according to Theophrastus, first put this question, which, separately from his *Natural history*, was remembered as *the* question of Xenophanes. Nicolaus of Damascus, the friend of King Herod, well known as philosopher too, has claimed that Xenophanes called the 'boundless' the 'beginning', whilst Theophrastus says that he called it the non-existent, 'having neither beginning, middle, nor ending'.[3] Parmenides, the pupil of Xenophanes, was faced with the same dilemma. For him the real world was not apparent, and the apparent world was not real.[4] It was only logical that a Sophist like Xeniades of Corinth, rejecting all thought of a 'real' world, which was not even apparent, should state boldly: The world is created from nothing; it is sham.[5]

It is here that we have to add and test Jewish and Christian testimonies for the creation from nothing. On the side of voices in favour of the reality of the world, although it is created from nothing, belongs an ancient credal formula of the Roman church, preserved by Hermas:[6]

[1] *Studia Theologica*, IV, Lund, 1950, 23 f.

[2] Atticus in Euseb., *Praep. Ev.* XV.6.7, ed. Mras, II, 360, 16 f. On Atticus cf. Überweg–Prächter, *Philos. d. Altertums*, 12th ed., 1926, 548 f.

[3] Theophr., *Natur. opin.* frg. 5, Diels, *Doxogr.*, 480 f.

[4] It follows from Theophr. frg. 6, Diels, *Doxogr.*, 482, that Parmenides' teaching was also widely remembered, but rather as a compromise.

[5] Cf. *Studia Theol.* IV, Lund, 1950, 25, and add to to note 3 there: Sext. Emp., *Pyrrhon. hypot.* II.18.

[6] Hermas, *Mand.* I.I.I. The fact that this was an early Roman symbol—or part of it?—follows not only from the introductory words, but even more so from Irenaeus, *Adv. haer.* I.15.1, ed. Harvey, I, 188 f., describing a very similar formula, '. . . one, almighty God . . . who created all things out of the non-existent', as *regula veritatis*.

'First of all you have to believe that there is one God, who has founded and organized the universe, and has brought the universe out of nothing into existence.' The Roman provenance of this formula is shown by the fact that it is used also by Justin Martyr and by Irenaeus.[1] However, the doctrine was by no means limited to Rome. It is also found in various non-Roman apologists, especially in Theophilus of Antioch.[2] Neither did it arise in Rome: it is easy at any rate to point to contemporary and earlier Jewish witnesses for it.[3] At the same time there was amongst the early Christians a strong inclination to characterize the empirical world as sham: 'The world is passing away with all its allurements', 1 John 2:17, was indeed firmly held, especially by the martyrs.[4] This conviction was based upon the doctrine that a world created from nothing would also be dissolved in nothing, was sham. The Maccabean mother had been the first martyr to put this point in her exhortation of her youngest son.[5] There is every reason to assume that the literary form of the Christian Acts of martyrs was dependent upon the Jewish, and especially the Maccabean model. They are, in spite of the many abuses hurled by the martyrs at the Roman magistrates sitting in judgment, very different from the pagan Acts of martyrs so-called.[6] Amongst the Christian Acts we find at least one, those of Marcianus and Nicander, where the Roman *praeses* exhorts the martyr Nicander: 'Honour the gods, if only with incense,' and receives the reply: 'How can a Christian worship wood and stones, and leave the immortal God who has made all from nothing?'[7] This source, which cannot be earlier than the middle of the

[1] Justin M., i. *Apol.* x; Iren., *Adv. haer.* I.15, loc. cit.; II.10.2; IV.34.2, ed. Harvey, I, 188 f., 274 f.; II, 213 f.

[2] Cf. e.g. Athenagoras, *De resurr.* 3; Theophil., *Ad Autol.* I.8; II.4; 10, ed. J. C. T. Otto, *Corpus Apol.* VIII, 1861, 26, 56, 78.

[3] Contemporary, cf. e.g. Apc. Baruch (syr.) 48:8, 'thou callest into life by thy word that which is not'. Earlier sources cf. M. Dibelius, *Handb. z. N.T.* Erg. Bd., 1923, 497.

[4] Cf. e.g. 'whilst they only seem to exist', Ignat., *Trall.* 10; 'nothing visible is good', Ignat., *Rom.* 3:3, cf. 2 Cor. 4:18; Ignat., *Rom.* 4:2 etc.

[5] 2 Macc. 7:28.—It is worth noticing that 4 Macc., which is soaked in Stoic philosophy, has omitted this exhortation, although it shows in 12:7, that an exhortation by the mother was found in its source.

[6] H. A. Musurillo, *The Acts of the pagan martyrs*, 1954, 238 f., seems mistaken in his attempt at equating even the Acts of the Alexandrian martyrs, edited by him, with those earlier ones of gymnosophist and Stoic martyrs. The most striking, and indeed perplexing, feature of the Alexandrian Acts is that they have never a reference to metaphysics.

[7] *Acta Marciani et Nicandri* I, ed. Th. Ruinaert, *Acta martyrum*, Ratisbonae, 1859, 571.

third century, is significant for the strength of a conviction which, because of the support it could lend to the docetic heterodoxy, was undoubtedly under suspicion.[1] At that time *creatio ex nihilo* vanished from the *regula fidei*.

IV

We have strayed on strange fields: can these be Greek ideas still? In answering this question, I do not believe that the inquiry is closed because their immediate sources have been shown to be Jewish. For even the Palestinian Jews were interested in the religious views of their non-Jewish neighbours; and the Jews of the dispersion were out to make converts, and many of them were converts themselves. Pagan convictions were certainly familiar to them, as St Paul's sermon before the Areopagus shows, whoever may have written it. Taking the Christian arguments in favour of creation from nothing separately, I feel strongly that the martyrs' representation of the empirical world as sham, combined with their hope of entering the 'real', eternal life, is very near to Eleatic teaching as described by Plutarch. The following is a verbatim quotation, substituting the one word 'kingdom' for universe:[2]

> Xenophanes left room neither for creation nor destruction, but claims that the *kingdom* (read: universe) is always the same. For if it were created there would have been a time when it was non-existent. But the non-existent cannot come into existence, neither can it make anything nor can anything be made from it. Consequently he calls the senses liars and he slanders together with them the logos, i.e. reason.

Apart from the last few words, where 'the so-called knowledge', 1 Tim. 6:20, would have to serve, 2 Cor. 4 and 5, and Colossians, would provide analogies not only to this report, but also to a related one about Metrodorus of Chius, a pupil of Democritus.[3]

The second point is the emphasis on monotheism. We find the Jewish formula εἷς θεός in the Roman credal formula of Hermas and of Irenaeus, but not, of course, in pagan writers. There was nevertheless, a

[1] Cf. the curious *sacrificium intellectus* by Iren., *Adv. haer.* v.3.2, ed. Harvey, II, 325 f., and Ps. Justin, *De resurr.* 5, ed. J. C. T. Otto, *Corpus Apol.* III, 1849, 222, both stressing man's creation from clay.

[2] Plutarch, *Strom.* IV, Diels, *Doxogr.*, 580.

[3] Plutarch, *Strom.* XI, Diels, *Doxogr.*, 582, cf. Seneca, *Epist.* LXXXVIII. 43, on another of Democritus' pupils, *Nausiphanes ait, ex his quae videntur esse, nihil magis esse quam non esse.*

strong henotheistic tradition in the Academy, the Peripatos, the Stoa, and particularly in Neo-Pythagoreanism. In the Academy there was the tradition of the eternal father, derived from the *Timaeus*,[1] in the Stoa we have seen the one deity surviving the conflagration. However, Hellenism being largely a conservative movement, found no difficulty in combining the one supreme god with a whole galaxy of deities.[2] We notice how Galenus recorded the fact that it had been the supreme deity who, according to Anaxagoras, created the cosmos, or according to Plato set it in order.[3] Of Aristotle it was reported that his supreme deity was detachable from the ball of the universe upon which it had taken its place. However, it had arranged this ball into spheres, and to each of these it had granted an eternal soul.[4] It is necessary to state here, and strongly, what classical scholars all over the world have failed to face: the idea of monotheism is Jewish. The religion of classical Greece never counted its gods, not even the 'twelve gods' of Olympia.[5] It was the personal God of the Jews who had to be one and only one. This bold challenge met with that emotional, sentimental religiosity so typical for European civilization ever since. The successful careerists of the early Roman Empire, like Cicero, were dreaming of a benevolent supreme providence which, we hear, every substandard intelligence will discover immediately as the creator of the universe by lifting its eyes to the starry sky.[6] The Christian apologists of the second century began the amalgamation of this sentimental religion with Jewish monotheism in early Catholicism, but it has proved the stronger of the two and has, for the last 250 years, figured as *the* Christian faith in popular conscience.

The third point was made most strongly by Irenaeus saying,[7] 'that the Father made all things himself . . . the God of Abraham, Isaac, and Jacob, above whom there is no other god, neither ἀρχή, nor power

[1] Plato, *Tim.*, 28C.

[2] Cf. esp. about the Neo-Pythagoreans O. Gilbert, *Griech. Religionsphilosophie*, 1911, 142.

[3] Cf. Galenus, *Hist. philos.* 35, Diels, *Doxogr.*, 618.

[4] Aëtius, *Plac.* I.7.32, Diels, *Doxogr.*, 305. We find the doctrine of the soul as ἀρχὴ τῆς φύσεως strongly revived in Neo-Platonism, e.g. Jamblichus.

[5] This has been precisely stated by Th. Zielinski, *The religion of ancient Greece*, 1926, 163 f., who contrasts this attitude not with Judaism, but with Islam, Anyone who has ever looked at the sources of E. Peterson, *Heis Theos*, 1926, with some understanding will know the greatness of Zielinski's mistake in this.

[6] Cf. e.g. Cicero, *De nat. deor.* II.97.

[7] Iren., *Adv. haer.* I.5.I, ed. Harvey, I, 188 f.

nor *pleroma*'. Undoubtedly this referred largely to the internal Christian conflict between Catholics and Gnostics; but this personification of the 'beginning' in the singular[1] reminds us of a saying by Galenus that 'some people believe that the ἀρχή was the guiding power in all things, not as the first cause of the whole creation, but preceding the causes in all things'.[2] These 'some people' were philosophers, although there is no indication to which school they belonged. However, the following Neo-Pythagorean fragment may well be said to contain a similar idea:[3]

It is clear that the ἀρχή by itself is uncreated and eternal, and the cause of creation and motion; and moving all things it is itself by itself, and producing the other absolute things, it is the absolute itself by itself. Analogies and similarities of this ἀρχή [may be found] in natural and technical production. Now since the immortal is the tireless, and the tireless the untiring mover; and since such a deity is ever moving the whole universe, it is clear that the cosmos is also eternal.

In short, the cosmos is eternal because of the immanence of its divine ruling principle. Yet, as Pythagoras himself was said to have held,[4] 'the cosmos was created according to design, but not in time'. Here, therefore, was a philosophy which Irenaeus' saying may have opposed. It was, however, not the only one. Pythagoras' formula was also used for Heraclitus,[5] of whom Galenus records that he described the immanent logos as 'the essence of fate (εἱμαρμένη)', and fate as 'the ethereal body and seed of the creation of the universe'.[6] It was here, of course, where philosophical and mystical doctrines met, and the Church had to be doubly watchful. Here the doctrine of the male monad and the female dyad as the beginning of the cosmos was domiciled, and could be used already by Xenocrates, Plato's pupil, to allegorize the union between Zeus and the cosmic soul,[7] and the arithmetical characteristics of odd and even numbers appeared as proof that ἀρχή and element were of necessity inseparable.[8] Perhaps the philosophers intended to curb the too-impetuous inventiveness of the mystics, but such efforts were in vain. It happened that popular superstitions were simply added, e.g. when the scholiast to Homer followed up his

[1] In the New Testament it is always the plural 'dominions' which is personified.
[2] Galenus, *Hist. philos.* 19, Diels, *Doxogr.*, 611.
[3] Ps. Aristaeus in Stob., *Ecl.* I.20.6, ed. Wachsmuth, I, 176 f.
[4] Aëtius, *Plac.* II.4.1, Diels, *Doxogr.*, 330b, 15 f.
[5] Aëtius, *Plac.* II.4.3, Diels, *Doxogr.*, 331b, 5 f.
[6] Galenus, *Hist. philos.* 42, Diels, *Doxogr.*, 620.
[7] Aëtius, *Plac.* I.7.30, Diels, *Doxogr.*, 304.
[8] Ps. Plutarch in Stob., *Ecl. de arithm.* 10, ed. Wachsmuth, I, 22.

quotation from Xenophanes with the remark that water was, of course, the element which destroyed souls;[1] it also happened that a great philosopher like Jamblichus made the obscure relations between the One, the whole, and the cause no less obscure by his profound Neo-Platonism.[2] Reading these oracular utterings creates an involuntary admiration for the rationalist pedantry of the anti-heretical authors, and their Greek ancestors.

[1] Homer, *Il.* VII.99 schol., ed. E. Maass, V, 1887, 239.
[2] Jambl. in Stob., *Ecl.* I.5.17, ed. Wachsmuth, I, 80.

10 Hellenistic creation myths

I

The last chapter has shown that the philosophical and mythological approaches to the problem of cosmogony lie very close together. On several occasions the barrier between them was not clearly marked, and a change from the one to the other unavoidable. This, I believe, was not caused by the strange mythologies breaking into the well-fenced preserves of Greek rational philosophy, but rather by the narrow boundaries of this philosophy. The realm within which it could be proved that an illogical statement was also an incorrect one did not coincide with the whole expanse of human thought. The demand for logical conclusiveness has led to sophistries, like that of Achilles and the tortoise, which had made *a priori* conclusions suspect; and the new scientific techniques of experiment and observation had invalidated numerous traditional convictions. This point was of a particular importance for the philosophical discussion of cosmogony. When Palaiphatus in the second century A.D. gave his assent to the thesis proposed by Melissus six centuries earlier, in 'the beginning is all that has come to be, and will be now', he made a confession of faith. For it is easy to make the claim that there is nothing new under the sun; but the reply that nobody will swim twice in the same river is superior logically as well as scientifically. However, whereas Aristotle had been justly proud of the progress achieved during the classical period of Greek philosophy by the use of scientific methods, the profound distrust in the validity of observed facts, voiced first by the Eleatics, had rapidly spread in Hellenistic times.

In any case, there could not be any observed facts when the creation of the universe was enquired into: 'Where wast thou when I laid the foundations of the earth?' was a question which would harass the honest Greek no less than the honest Jew.[1] There was no choice here but to renounce the basis of experience, and to start from outside the world of sense perception. In this situation Hellenistic philosophers revived the memory of two closely related religious and philosophical movements

[1] Job 38:4, cf. Xenophanes frg. B 34, Diels–Kranz, 5th ed., I, 137, 2 f., 'for no man sees what is certain, nor will there be anyone who knows about the gods and what I say about the universe. Even if he happens to speak most perfectly, nevertheless he knows nothing: Mere opinion is the seal upon everything.'

which had taken the same stand, Pythagoreanism and Orphism. Pythagoreanism had undoubtedly discovered a realm outside the empirical world, the realm of pure mathematics; and Orphism had claimed to have done the same with the realm of the dead. This claim was strongly supported by the Pythagorean achievement. The Pythagorean conviction that the immaterial cosmos of numbers and geometrical figures was the principle of the empirical world, by which the validity of empirical observations could be tested, had proved itself. It had also become known that mathematical doctrines were universally applicable, and equally convincing in Egypt and Mesopotamia, and even in mysterious India, as in Italy and Greece. They could even be demonstrated by the movements of the celestial bodies. The general veneration for the celestial bodies, which could not be influenced by human power, but whose movements were predictable by human reason, made the assertion credible that this human reason was akin to the power that moved the heavens, was divine. 'As some of your poets have said; "We are also his offspring."'[1] This claim to divinity had probably caused the original *rapprochement* between Pythagoreanism and Orphism in the classical period, although their conceptions of immortality were not identical. Men like Pythagoras and Empedocles claimed to be divine as of right; the Orphics, however, obtained their eternal bliss by initiation. In Hellenistic time Pythagoreanism was revived, but roused the suspicion of the Philistines to such an extent the Roman police described

[1] It is annoying to read in E. Hänchen, *Die Apostelgeschichte*, 11th ed., 1957, 464 n.4, the peremptory verdict, 'Bruce, 338, thinks wrongly of Epimenides'. F. F. Bruce, *The Acts of the Apostles*, 2nd ed., 1952, 338, has not only made it probable that Theodore of Mopsuestia referred the words, 'for in him we live and move and have our being', to a panegyric of Minos over his father Zeus, but also that this panegyric started with the line, 'Cretans, always liars, evil beasts, slow bellies', quoted in Titus 1:12, which was ascribed by Clement of Alexandria to Epimenides, one of the semi-mythical Orphic worthies. In the Syriac translation the line introduces a quatrain which Bruce, following a probable suggestion by Rendel Harris, ascribes *in toto* to Epimenides. The last line of it is virtually identical with Acts 17:28. It is also a characteristically Orphic view, a fact which supports the ascription, cf. O. Kern, *Orphic Fragm.*, 1922, frgg. 167–9. All these considerations are dictatorially dismissed by Hänchen without any reason or word of discussion. It seems to me, however, that Epimenides might even be included among the 'some of your poets' since his saying, frg. B 2, Diels–Kranz, 5th ed., I, 33, 1, *καὶ γὰρ ἐγὼ γένος εἰμὶ Σελήνης ἠυκόμοιο*, shows a verbal identity with Aratus, *Phaen.* 5, and other parallels as Pindar, *Nem.* VI.1 f.; and since Cleanthes, in *Jov.* 4 f.; Cic., *De nat. deor.* I.91, illustrate the fact that this quasi-biological kinship between god and men was a legacy from Orphism to Stoicism.

all magicians as *mathematici*.[1] In Stoicism, however, various Orphic and Pythagorean tenets were adopted already by its founders, perhaps because they recognized an old kinship between those Greek esoteric doctrines and similar Near-Eastern beliefs, in order to claim more impressively that all human souls were divine.[2]

Man's 'real' being was in Hellenistic time placed in his soul; and this brought about in due course a great change in his ontological approach to the empirical world. For the soul, according to Orphic belief, did not belong to the empirical world,[3] which is of course not identical with the material world of Stoics and Epicureans. This Hellenistic change took its origin from Plato's idealistic philosophy; but it must not be forgotten that the group of pre-Socratic philosophers whom we have discussed had not been purely natural philosophers either, not even Anaxagoras.[4] If the word 'metaphysics' was of a later coinage, the philosophies of Anaximander, the Pythagoreans, Empedocles, and the Eleatics, all made allowance for the influence of that which was beyond nature upon their explanation of the empirical world. This influence, however, could vary in kind very much. On the one hand, there were the cosmogonic myths in Hesiod, the Orphics, and presumably a large hieratic tradition now lost. It is not our task to enquire into their origins.[5] Their material may have been taken from Babylonian or Egyptian sources, or may even be genuinely Greek; what has to be realized is that these myths exercised authority because they contained revealed truths, *ἀλήθεια*, in the truly Greek sense of 'that which is not hidden'.

Such a revelation, however, could take place in two ways. The light could be made to shine upon objects that were normally covered by darkness in an otherwise perceptible, visible world; and these objects would then be explained by a deity in their relation to, and significance

[1] The term occurs first in Tacitus, *Ann.* II.32; Juvenal VI.560; XIV.248, and often in Suetonius, and thus shows itself as belonging to the period of Domitian. From the same time onward it was technically used in legal language, cf. E. Massoneau, *La magie dans l'antiquité romaine*, 1934, 171 f.

[2] Cf. e.g. W. Jaeger, *Aristotle*, transl. R. Robinson, 2nd ed., 1948, 22.

[3] So the doctrine of Posidonius, cf. M. Pohlenz, *Die Stoa* I, 1948, 229 f., which was an elaboration of earlier Stoic doctrines, cf. ibid., 92 f.

[4] J. Bernays, *Abh. Akad. Berlin* 1882/3, 8 n.3, has made this point strongly, referring to Anaxagoras frg. B 12, Diels–Kranz, 5th ed., II, 38, 2, cf. frgg. A 15, 55, 100, ibid. II, 10, 22; 20, 32; 29, 30.

[5] I still read with admiration the chapter on cosmogony in J. E. Harrison, *Prolegomena*, 2nd ed., 1908, 624 f.

for, the known part of the world, a revelation of something new, an objective revelation. Or else it may consist in no more than the removal of a veil, or tinted glasses, from the seer's eyes by liberating him from impurity or injustice which obscured his sight, subjective revelation as in the case of Empedocles. Objective revelation increases the knowledge of the seer by increasing his experience; subjective revelation enables the seer to gather correct and relevant experience by his own efforts. Objective revelation was granted e.g. to Parmenides at his interview with Dike at the gates of day and night.[1] Empedocles, on the other hand, had to go through a process of purification which, I believe, was the necessary preparation for his work *On nature*. For it restored him to his rightful position as god. He had never ceased to be a god, different from ordinary humans,[2] but had been incapacitated by his defilement from gaining divine knowledge. That veil removed, he was himself in the place of the creator, entitled to create nature in his own mind. The objective revelation, however, involved a sort of spiritual promotion. We do not know how Parmenides was chosen to be transported on the solar chariot to the place of revelation; but however extraordinary a person he may have been, he was human, and whatever promotion he may have received at his revelation, it was by grace, and not a restitution to his rightful place. The two philosophers represent two types which should be distinguished amongst those people who spread esoteric doctrines in Hellenistic time: Empedocles as the magician, and Parmenides as the initiate.

II

There are various sources for the appreciation of the magical approach to the problem of creation in Hellenistic time; but it has to be stated with regret that research in these sources has often suffered from methodological vagueness. It is neither safe, nor even particularly interesting, to state that those magical texts which have been found on Egyptian papyri[3] are strongly influenced by the national religion of Egypt; or

[1] Cf. my little study 'Dike am Tor des Hades', *Studi in mem. E. Albertario* II, 1950, 547 f. [2] Cf. K. Kerenyi, *Pythagoras und Orpheus*, 3rd ed., 1950, 24.

[3] A considerable proportion of these texts have been published most carefully by K. Preisendanz, *Papyri Graecae magicae* I, 1928; II, 1931. A third volume, containing most of the smaller items, and the indexes, was announced, but has never appeared. The papyri from this collection will be subsequently referred to as *PGM*. As regards the origins of the religious beliefs expressed in these papyri, Greek ideas about the ancient Egyptian religion, as found in the texts in Th. Hopfner, *Fontes hist. relig. Aegypt.*, 1922–5, are prominent, as are also Jewish influences (cf. E. Peter-

that others which contain the names of 'Chaldean' wizards so-called, like Ostanes and Zoroaster,[1] are testimonies for Iranian beliefs, even if a considerable minority of them reflect the fundamental spirit–matter dualism which, on the whole, is absent in Greek classical thought. The Greek magical papyri, upon which I have concentrated my research, are Greek in the sense that they are representative for the Greek-inspired Hellenistic religious syncretism in the Roman Empire. Limiting my remarks to them has the advantage of dealing with a large, homogeneous group of documents. They mostly belong to the century from A.D. 250 to 350, and come all from Egypt. They are, however, representative for a much longer time, and a much larger part of the world, as comparisons with Latin magical texts, the so-called *tabulae defixionum*, have shown me.[2] These documents show in many cases a similar technique in the magic practices, as does also the *Apologia* of Apuleius.[3] It is also alleged that these papyri are only representative for the vulgar superstitions of the mob. This view is only partly true. The actual papyrus codices are carefully written, and modern research has shown that papyrus, even in Egypt, was not cheap.[4] But even if these codices were of humble origin, some of their contents are not. The great magic centres in Egypt were Memphis, Heliopolis and Hermopolis, and it is well known that the Emperor Hadrian, on his journey through Egypt in A.D. 130, was in the two latter places initiated to their 'sacred magic', and that the death of his favourite Antinous at Hermopolis on that journey was presumably due to apotropaic magic.[5] It is, therefore, significant that in the majority of cases Helios or Hermes are the deities invoked, that one of the papyri has preserved a fairly long passage of the Hermetic tract *Asclepius*,[6] and that Hadrian's initiation is expressly men-

son, *Frühkirche, Judentum und Gnosis*, 1959, 107 f.) and Hermetic influences. Obviously, there are also strong 'Chaldean' influences; but it seems a mistake to regard a whole as no more than the sum of its parts.

[1] The collection of sources by J. Bidez–F. Cumont, *Les mages hellénisés* I/II, 1938, has its particular value in the analyses of the individual sources.

[2] Some of these have been carefully edited and annotated in V. Arangio-Ruiz & Olivieri, *Inscript. Graecae Siciliae et infimae Italiae*, 1925, 157 f.

[3] F. Norden, *Apuleius v. Madaura u.d. röm. Privatrecht*, 1912, 29 f., gives a convenient list made by the prosecution.

[4] F. Wieacker, *Textstufen röm. Juristen*, 1960, 97 n.22.

[5] Cf. F. Beaujeu, *La religion romaine à l'apogée de l'empire* I, 1955, 243.

[6] P. Mimaut = *PGM* IV. 591 f., Preisendanz I, 56 f. is the Greek of Asclepius 41 fin., Nock–Festugière, II, 359 f. On further literary connections with Lucian, *Philopseudes*, cf. R. Reitzenstein, *Hellenist. Wundererzählungen*, 1906, 5 n.2.

tioned in another of these papyri.[1] These are reasons for seeing much of the material contained in the papyri in a close connection with those magic centres.

Not only the quotation from *Asclepius*, which is of second-century origin, shows a superior style; but the same may be said of most of the invocations of such deities as Helios, Hermes, Apollo, and Aphrodite, and such invocations are both numerous and elaborate. They are also of a considerably greater age than the manuscripts in which chance has preserved them. Their literary origin, however, is uncertain, and especially whether they are plagiarisms from pagan liturgies.[2] For it has to be considered that pagan religious literature, which presumably was produced in order to compete with Jewish and Christian writings, was more substantial as well as more easily accessible than the pagan liturgies, the loss of which is so widely deplored. Some of these invocations, it is true, show a hymnic character, but even the use of hymns is not uncommon in contemporary religious literature. They are to be found in Hermetic[3] as well as in Christian[4] tracts; and there exist also several independent collections of hymns of Orphic, Essene, and Manichaic origin, as well as the Christian *Odes of Solomon*. On the existing evidence, taking into consideration the proved quotation from the Hermetic *Asclepius*, a non-liturgical origin of most of these magical invocations seems likely.

As a whole such a magic papyrus presents an unwholesome mixture of human, all too human, desires; meaningless so-called magic words; fanciful, sometimes disgusting magic actions and recipes; and finally invocations and prayers, which stand out by their higher stylistic level.[5] All magic invocations, and ours are no exception from the rule, follow a line of combined blackmail and flattery. The principle is to make the invoked deity aware of the fact that not only its power but also its

[1] *PGM* IV.2447, Preisendanz, I, 149.

[2] Thirty years ago scholars were still very confidently affirming this, cf. F. Cumont, *Die orientalischen Religionen*, transl. Burckhardt–Brandenberg, 3rd ed., 1931, 212 n.8, with references to other literature.

[3] Cf. the hymn at the so-called consecration of the prophet, *C.H.* XIII.17–20, Nock–Festugière, II, 207, where the injunction of silence upon nature at the beginning of the hymn has several parallels in the magic texts, e.g. *PGM* I.198 f., Preisendanz, I, 40.

[4] Cf. the hymn at the end of Clem. Alex., *Paedag.*, ed. O. Stählin, I, 291 f.

[5] Cf. A. Dieterich, *Eine Mithras-Liturgie*, 1903, 31 f.

violability is known to the magician. He therefore makes constant use of the formula 'I am', which is so well known to us from biblical literature, and especially the Fourth Gospel; and it is by no means a stereotyped formula.[1] On the contrary, these statements are at times unbelievably strong: 'I am he who thrones on the two Cherubim in the midst of the two natures, heaven and earth, sun and moon, light and darkness, rivers and sea', says such a warlock, and with these words he conjures up 'the archangel from beneath the cosmos, the ruler Helios, who is subject to the one and only one. The eternal and only one commands it.'[2] Such a command went out into all nature: 'Listen form and spirit, and earth and sea, to the speech of the wizard of Ananke (Necessity), and take notice of my words as of fiery darts, for I am the man of god in heaven.'[3] Another magician uses a more courteous form: 'I invoke thy holy and great and secret names in which thou rejoicest';[4] and at times even an almost devout approach is made: 'I have shouted thy unsurpassed glory, creator of gods and archangels and decani.'[5] In all these invocations it is presupposed that the god is desperately keen on conversing with the magician, but that his purpose in doing so is uncertain.

In the last invocation the title of the deity invoked as the creator has been quoted here on purpose. For similar titles are used in all the other invocations which have been mentioned, as well as in several others:[6]

Come to me in the sacred circulation of thy holy spirit, creator of the universe, god of gods, lord of all, who hast distributed the cosmos with thy divine spirit. Thou hast appeared first from the first-born cruel water, begotten in a miracu-

---

[1] E. Schweizer, *Ego eimi*, 1939, 29 f., who maintains this, cannot have read the *PGM*.

[2] From the 2nd Leyden pap., *PGM* XIII.254 f., Preisendanz, II, 100.

[3] *PGM* IV.1174 f., Preisendanz, I, 112. E. Peterson, *Frühkirche, Judentum, Gnosis*, 1959, 107 f., has given a learned, if venturesome, interpretation.

[4] *PGM* IV.1609 f., Preisendanz, I, 124.

[5] The whole passage, *PGM* IV.1189–1207, Preisendanz, I, 112, where these words are found, is interesting because it shows the degradation of texts of higher magic for vulgar use by the interpolation of the words: 'Guard me, the N., from proud officialdom and all insults, or do such and such a thing, lord, god of gods (magic word), creator of the cosmos, creator of all things, lord, god of gods.' Further significant 'I am' formulae may be found in *PGM* V.145 f., 476 f.; XIII.334 etc.

[6] Similar prayers appear frequently also in Orphic hymns, where the title of creator, given to Night, Heaven, Pan, Heracles (!), Kronos, Rhea, Zeus, and Proteus, but not to Apollo, Hermes, Dionysus, Eros, may have a connection with the earlier creation myths.

lous way, thou who hast created the universe, the abyss, earth, fire, air, and again the ether, the roaring rivers, the red-shining moon, and the stars high in the air, matutinal, roving. To thy counsels they submit in everything.[1]

Such descriptions of the creative power of the deity are frequent in the magical papyri, and characteristic for them.[2] They may appear in metrical form;[3] they may even mention a creation from nothing.[4] The literary type is closely related to the technical, one may say professional, praise of the great deeds of the godhead called aretalogy,[5] and they should be seen as shortened aretalogies. If this is clearly understood it will not come as a surprise that in two cases this type of invocation has been enlarged to a complete aretalogy. It consists of a creation myth in two different, but related versions, the magical creation myth.[6] The shorter version runs as follows:

So he spoke and clapped his hands. And the god laughed seven times; and as the god laughed there came forth seven gods who surround the cosmos. For these are the first that appeared.

When he chuckled first light appeared and brightened the universe. He chuckled the second time, and all was water. Earth hearing the noise[7] cried and writhed, and the water was trisected. A god appeared; he was placed over the abyss. For without him the dampness neither rises nor falls. But when he wanted

---

[1] *PGM* III.549 f., Preisendanz, I, 56.

[2] Cf. e.g. *PGM* I.29 f.; IV.1137, Preisendanz, I, 4, 110, preceded by ἀρχὴ καὶ τέλος τῆς ἀκινήτου φύσεως, line 1125; IV.1749 f., where Helios is described as the ἀρχηγέτης πάσης κτίσεως.

[3] Cf. the hexameters in *PGM* IV. 1957–1989, Preisendanz, I, 132, repeated in an inferior tradition in *PGM* VIII.74f., Preisendanz, II, 49.

[4] *PGM* IV3075; XIII.271 f., Preisendanz, I, 172; II, 101. The former passage, mentioning in its introduction 'Jesus, the god of the Hebrews', shows strong gnostic influence, the latter passage, invoking Thayth, *Hermetic influence*, cf. *C.H.* frg. IX, Nock–Festugière, III, 51.

[5] Cf. R. Reitzenstein, *Hellenistische Wundererzählungen*, 1906, 8 f., and notice how the magician in *PGM* XIII.445, claims to be appointed (ταχθείς) by Helios to proclaim his miraculous deeds.

[6] *PGM* XIII.62 f., 443 f., Preisendanz, II, 90 f., 109 f. The longer, second version evens out various objectionable points in the first. They are, however, both dependent on the same prototype, and may have circulated separately. They were each received to an 8th book of Moses, which were combined in *PGM* XIII, cf. R. Reitzenstein, *Die Göttin Psyche*, S. B. Heidelberg, 1917, 28 f. Jewish origin, suggested by Ad. Jacoby in Preisendanz, II, 95 n.1, is unlikely. The trisection of the water, adduced as evidence, has its root in 'Egyptian' belief in a sub-tellurian ocean, cf. Servius ad Vergil, *Georg.* III.363, ed. Thilo, III, 1887, 347.

[7] Cf. the irrational outcry of the waters in *Poimandres* 5 fin., Nock–Festugière, I, 8, 3 f.

to chuckle the third time, because of the god's bitterness there appeared Nous, holding a heart, and he was called Hermes.[1] The god chuckled a fourth time, and there appeared procreation (*Γέννα*) holding the seed. The god laughed for the fifth time, and laughing he was sad. And there appeared fate (*Μοῖρα*) holding scales, indicating that in her was justice. But Hermes quarrelled with her, saying: 'Justice is in me.' Whilst they fought the god said: 'Out of both of you justice shall appear; and all things in the cosmos shall be subject to thee.' And she took the sceptre of the cosmos first. And the sixth time he chuckled and rejoiced greatly. And there appeared opportunity (*Καιρός*) holding a sceptre, and handed the sceptre to the first-created god. And taking it he said: 'Thou shalt be the next after me, carrying the glory of the light.' When the god chuckled the seventh time, Psyche was made, and chuckling he wept. Seeing Psyche he whistled, and earth writhed and brought forth the Pythian dragon, who knows all things in advance. Seeing the dragon the god was shaken and clicked his tongue; and by the god's clicking an armed man appeared. Seeing him the god was frightened again at the view of a stronger one, lest Earth had thrown out a god.[2] Looking down upon earth he said: Iao; and by the sound a god was born, who is lord of all. But the former clashed with him, saying: 'I am stronger than this one.' But the god said to the strong one: 'Thou hast happened by my clicking, but he by sound; you two will be together over all necessity.'

It may well be held that this creator god was not so very omnipotent; but the question remains, where did the magician's arrogance come from with which he added, 'lord, I imitate thee with the seven vowels, enter and give ear to me'?[3]

It was in this connection that the term ἀρχή entered the magic realm. The magician invoked the god as 'first creation of my creation (seven vowels), first beginning of my beginning (whistle three times, click thy tongue thrice, blow), first spirit of my spirit (roar thrice), first fire in my composition, god-given among the components, the first of the fire in me etc.'[4] Although this invocation has been preserved in a dif-

[1] Cf. *C.H.* VII.2, Nock–Festugière, I, 81, 17, saying that Hermes can only be known νῷ καὶ καρδίᾳ. It may well have been the bitterness of tract VII, which caused the bitterness of the god.

[2] In the gnostic tract 'On the archons', German transl. in Leipoldt–Schenke, *Kopt.-gnostische Schriften*, 1960, 73, 14 f., the spirit 'comes forth from the adamantine earth'.

[3] It is significant that an attempt to adapt the seven vowels to the Hebrew names of the seven sons of Jaldabaoth seems to have been made in the Ophite tract 'On the origin of the cosmos', German transl. by H.-M. Schenke, *Theol. Lit. Z.*, 1959, 249 f.

[4] *PGM* IV.488 f., Preisendanz, I, 88 f. The whole passage, 475–535, is of great importance for the concept of theurgic magic, and by no means to be seen as vulgar sorcery.

ferent papyrus codex, the animal noises[1] and the seven vowels bring back the atmosphere of the magical creation myth. The magician claims his kinship with the god. We also possess an elaborate description of what constitutes the magician's ἀρχή. For under this heading a kind of mystical union between him and the deity is described in the following terms:

Opened were the gates of heaven, opened the gates of the earth, opened was the path of the sea, opened the path of the river. My spirit was heard by the spirit of heaven, my spirit was heard by the spirit on earth, my spirit was heard by the spirit of the sea, my spirit was heard by the spirit of the rivers. Therefore grant the spirit to the mystery that I have prepared, ye gods whom I have invoked by their names, grant breath to the mystery that I have prepared.[2]

This invocation has been wrongly characterized as an initiation or consecration by the editor. It is nothing of the kind. It is another 'I am' formula, the opening of a theurgic operation, but this time intended for general use, so to speak the magician's passport to the divine realm. The magician claims that he is not barred from the councils of the gods: For he himself is a god.[3]

III

It was not only the magicians who believed in their own deification. The gnostics too held such a hope; and it is perhaps no basic distinction that the more considerate amongst them expected it only after the completion of earthly life, and saw their exalted position on earth, as the bearers of true gnosis, as a provisional and conditional one.[4] It is therefore not surprising to find that certain gnostic documents exhibit an opinion on the creation of the world, closely related to the magic creation myth. There are in particular two tracts from the gnostic library found near Nag-Hammadi, which have to be considered here: *On the archons*, which has been published in full; and *On the origin of the*

[1] They are prescribed again in the same ritual, *PGM* IV.578 f., Preisendanz, I, 92.

[2] *PGM* XII.323 f., Preisendanz, II, 80.

[3] Cf. the *Gospel of Philip* 44, German transl. in Leipoldt–Schenke, *Kopt.-gnost. Schriften*, 1960, 46, 'you have seen the spirit, and you became spirit; you saw Christ, and you became Christ; you saw the Father, and you shall be the Father'.

[4] Cf. the *Gospel of Philip* 4, Leipoldt–Schenke, 39, 'he who has come to believe in the truth, has found life; and he is in danger of dying. For he lives.' Cf. also the provisions made for the recovery of the lapsed initiate in *Pistis Sophia* 105, edd. C. Schmidt–W. Till, 172, 21 f.

world, the first half of which is now at our disposal.[1] Both are products of the Ophite school, which was described already by the early Catholic fathers as singularly open to non-Christian influences.[2] In addition they may be described as not a little obscene. The most important point about the magical cosmogony is that it retails two creations; and that the god who 'laughed seven times' is not identical with the 'chuckling' god.[3] This second god, whom we have described as not very omnipotent, is clearly the Jaldabaoth of the Ophite myth; and the world which he creates is below the eternal Ogdoad, and apparently is meant as the sub-lunar world.[4] This world was formed from—and presumably in—the dark chaos of matter. There existed originally a strongly dualistic tendency in gnosticism first formulated, so it seems, by Basilides.[5] Marcion too taught that the god of the Jews had created the world from matter, and that matter was evil.[6] Our Ophite myth, however, is a compromise, even more so than the related one in Pistis Sophia.[7] There are, however, two characteristic features in our Ophite myth. The first is that it sets out to prove that chaos and matter are not uncreated, eternal, but have an origin in time. They are in fact the shadow of a 'work' of Pistis Sophia, which seems to be the heavenly cross.[8] The second follows from the first. It is that this Pistis Sophia was not trapped in the world of matter by an evil demon, but acted as mediator between the first Ogdoad and the world of matter in a similar way as the 'seventh power', the junior spirit, in the Simonian system.[9]

In this world of matter and chaos the original ruler was Jaldabaoth, the abortion of Pistis Sophia, a blind god because he could not and would not recognize the eternal light. With matter he produced seven sons, the seven planets; and in their presence he insulted the highest god by claiming that he alone was god. For this crime he was thrown into Tartarus. However, by the intervention of Pistis Sophia and her daughter

[1] The first tract is contained in Leipoldt–Schanke, 69 f.; the second in *Theol. Lit. Z.*, 1959, 249 f., both in German translation.

[2] Cf. the quotation from Hippolytus of Rome in W. W. Harvey, *Irenaeus* I, lxx n. 2.

[3] R. Reitzenstein, *Die Göttin Psyche*, S. B. Heidelberg, 1917, 32 f., neglects this completely.

[4] W. Bousset, *Hauptprobleme der Gnosis*, 1907, 9 f., did not see that the gnostics added the world of the seven planets to the sub-lunar world of chaos.

[5] Cf. W. Bousset, op. cit., 92 f.

[6] A. Harnack, *Marcion*, 2nd ed., 1924, 97 f.

[7] Cf. W. Bousset, op. cit., 99 f.

[8] On the heavenly cross cf. *Acta Johannis* 99, ed. M. Bonnet, II.1, 201 f.

[9] Hippolytus, *Refut.* VI.14.3 f.

Zoë some sort of existence was granted to this world. Sabaoth, one of the sons of Jaldabaoth, whom Bousset identified with the planet Mars, but who in this context at any rate is the sun,[1] received mercy because of his repentance. Joined to Zoë as his companion, he achieves an almost legitimate rule as 'lord of the powers' in the sub-lunar world.

It will be evident that this gnostic myth has little in common with biblical Judaism, and nothing at all with New Testament Christianity. However, it was later linked somewhat more closely with Christian teaching by the Valentinians. This gnostic sect possessed the Fourth Gospel, and reverenced it. They accordingly used its prologue in order to prove that the gnostic cosmology was scriptural. Irenaeus has preserved for us the following passage from a Valentinian treatise, which is typical for their argumentation:[2]

John, the disciple of the Lord, intending to speak about the genesis of the universe, by which the father brought forth all things, proposes that a 'beginning' was the first product of god, which he calls the son, and the only begotten, and god, in whom the father placed all things as seeds. By him, he says, the word was brought forth, and in this the whole essence of the aeons to which the word afterwards gave form. As he speaks of the first genesis, he does well to start his teaching with the 'beginning', i.e. the son and the word.

For he says: 'In the beginning was the word, and the word was with god, and the word was god. The same was in the beginning with god.' First he separates the three, god, the beginning, and the word; then he unites them in order to show the production of the two, the son and the word, as well as their union with each other and with the father. For in the father and from the father is the 'beginning'; and in the beginning is the word. So he states well, 'in the beginning was the word', for it was in the son. 'And the word was with god', for the beginning[3] . . . 'and the word was god' accordingly; for that which is born of god is god. 'The same was in the beginning with god', shows the order of production. 'All things were made by him, and without him not a thing was made': For the word became the cause of the formation and creation of all aeons after it. But he says, what was made in him was life, and here he signifies

---

[1] W. Bousset, *Hauptprobleme*, 10 n.1, 2, quotes Mandaic parallels in support of his claim that Adonaios represented the sun in the Ophite system. However, Origen has omitted Adonaios completely, and it is mere conjecture on the part of Bousset that he ought to be given the place of the sun in Origen's list.

[2] Iren., *Adv. haer.* I.8.5 f. Harvey, I, 75 f. F. M. M. Sagnard, *La gnose valentinienne*, 1947, 306 f., ascribes the text to Ptolemy, which is not certain, but probable.

[3] Here the verb is missing, and in spite of Harvey's attempt at justifying this omission, I agree with W. Völker, *Quellen zur Geschichte der christlichen Gnosis*, 1933, 94, that we have to find here a very old gap.

the syzygy (the companionship). For the universe, he says, was made by him, but life in him. Thus that which came to be in him is more akin to him than that which was made by him. For it is united with him, and bears fruit of him. For he continues: 'And the life was the light of men'; using just man, he signified the church together with [original] man, so as to make clear by the one term the unity of their companionship. For out of Logos and Zoë are Anthropos and Ecclesia etc.

This goes on for quite a while in a hybrid style combining philosophical and mystery thought-forms, till the conclusion is reached: 'Thus spoke John about the first Ogdoad, the mother of the aeons. For he spoke of the father, Charis, Monogenes, Aletheia, Logos, Zoë, Anthropos, and Ecclesia.'[1]

Whilst it is a sobering thought that this passage is a piece of the scientific theology of the second century, a discipline which we see artificially revived today by our scientific theologians, it is also a wholesome one. Admittedly, their aims are different. Our modern demythologizers need a myth, of course, in order to demythologize it;[2] the ancient Valentinians needed it in order to accommodate their preconceived ideas of Iranian dualistic origin within the Christian gospel. The possibility that there might be no myth at all is proscribed as unscientific by both. The gnostic and the magic myths seem to have sprung from the very same root. However, in the circles of the magicians interest was concentrated upon the sub-lunar world, even if their myth should have originally described the creation of a 'Chaldean' cosmos, consisting of a

[1] Cf. the comments by F. M. M. Sagnard, op. cit., 311 f., who, however, isolates the passage too much from the general gnostic tradition. It is hardly correct to say, 'la gnose ira jusqu'à identifier matière et mal'. It is rather true to say that the gnostics never fully succeeded in freeing themselves from the Iranian dualistic conviction that matter was evil.

[2] R. Bultmann's 'non-Faustic' speculations, *Das Evangelium des Johannes*, 10th ed., 1941, 6 f., are invalid, unless he can show a pre-Christian gnosis with the god Logos as *the*, or at least *a*, central figure, which is compatible with Iranian dualism. Bultmann's claim that *ἀρχή* in John 1:1 must be 'before all time and all the world' is theologically correct; but on his hermeneutic principles it is contrary to the evidence. In the Pre-Socratics and Plato, and even in Aristotle, such an idea is to be found. It vanished in Hellenistic thought, and is difficult to establish even in the Fourth Gospel. John 17:24, quoted by Bultmann, means it, if it is disregarded that for the gnostics the *καταβολὴ τοῦ κόσμου* was the 'second' creation, which took place in time. In *C.H.* IX.6, Nock–Festugière, I, 99, 5 f., it is made clear that the simile was taken from sowing, obviously in due season. It seems, therefore, that on his hermeneutic principles Bultmann contradicts himself when he disregards these analogies.

firmament and seven planetary spheres. And the throne of the ancient creator-god showed itself tottering in the magic myth, which was anything rather than a truly sacred story.[1] The Ophites resolutely sent him to hell, and his world into the chaos of matter.

It seems, however, that in the Hermetic texts the ancient régime is represented in a healthier state.[2] The Hermetic myth was, of course, deeply influenced by Greek philosophical ideas. As a significant example I would mention the repeated references to the old doctrine of *ἓν τὸ πᾶν* in the Hermetic writings.[3] This doctrine became popular in Hermetism because of its close alliance with the Alexandrian aeon-doctrine;[4] but the works of the Eleatic school, whose watchword this maxim had been, were also still available in the libraries of Alexandria.[5] Unfortunately the texts in our Corpus Hermeticum are without exception of a late origin, and in particular those on cosmogony. We find e.g. in one of the significant passages which stress the identity of god, the good, and the 'beginning',[6] an attack upon the idea of a suffering god, which evidently was a scarcely veiled attack upon Christianity.[7] (An open attack was out of the question since the revelations of Hermes were meant to be timeless, whilst Christ's passion had taken place in recent time.) This places its time of origin at the beginning of the third century. Among these texts of a late origin is a treatise called 'The virgin of the world', of which large fragments were preserved by the Byzantine collector John Stobaeus. The first of these fragments contains a creation myth, which has been ascribed to the pre-Christian period.[8] There can be no doubt that Plato's *Timaeus* was one of its ancestors, but this fact alone is not enough for its explanation. For the myth consists of several different components, and the reason for adducing it here is that another

[1] R. Reitzenstein, *Hellenist. Wundererzählungen*, 1906, 7, rightly stresses that Odysseus was the father of aretalogy. *Mendax aretalogus*, was Juvenal's, *Sat.* XV.16, verdict.

[2] R. Reitzenstein, *Die Göttin Psyche*, S.B. Heidelberg, 1917, 38 f., has already stated this; but he has paid no attention to the evident signs of decline and decay in the magic myth.

[3] Cf. *C.H.* V.9 and 11; XI.20, with Reitzenstein's remarks, *Poimandres*, 239, 243; Asclepius 2, 3, 10, 20; and in particular *C.H.* X.14, with the comments in Nock-Festugière, I, 129 f. n.51.

[4] Cf. the vignette on the title-page of E. Norden, *Agnostos Theos*, 2nd ed., 1929.

[5] J. Bernays, *Abh. Berlin* 1882/3, 12 f., has made this claim strongly on behalf of *De aeternitate mundi*.

[6] Cf. e.g. *C.H.* III.1; IV.10; VIII.5, Nock–Festugière, I, 44; 53, 1; 39, 9 f.

[7] *C.H.* VI.1, Nock–Festugière, I, 72, 3 f.

[8] Cf. E. Norden, *Agnostos Theos*, 2nd ed., 1929, 65.

Hermetic creation myth has been shown to be very near to the magic creation myth, but without throwing much light on its earlier stages.[1] Assuming then that the 'Virgin of the world' belonged to the same, Egyptian, milieu as the *Poimandres* which, I think, is likely,[2] its creation myth should be close to the magic myth as well. It is an aretalogy in type too, and its spirit is akin to the magical philosophy, the prototype of which we have found in Empedocles. For in the 'Virgin of the world' we receive a general instruction about those select souls which are destined to be kings, prophets and seers among men, lions among beasts, eagles among birds, or dolphins among fishes.[3] More important is it for us that the pre-existent god in the Hermetic myth is challenged to create, and smiles in reply.[2] Granted he does not laugh; but that is a Homeric retouch which is germane to the literary type of aretalogy: Zeus also smiled when replying to a request made by one of his lady daughters.[5] It is necessary to state that a smile at the beginning of the creation of this world of tears was not a matter of course.[6] The smile of the divine child was regarded as the most propitious omen for the new aeon which it inaugurated;[7] the laughing or otherwise of the creator are the *omina* for his world. This, I believe, is the decisive common feature of the two myths. They also have in common the two stages of creation; the rise of the strong one, Momus, from earth, and his discussion, though not an open quarrel, with Hermes;[8] and finally the creation of the tellurian gods by the sound of the creator's voice.[9]

[1] Cf. R. Reitzenstein, *Die Göttin Psyche*, S.B. Heidelberg, 1917, 33 f., comparing the magical myth with *Poimandres* 4–5, Nock–Festugière, I, 7 f.

[2] Nobody, I believe, has so far doubted this; but I feel that the local material from inscriptions etc., which was used by me in 'Lass die Toten ihre Toten begraben', *Studia Theol.* VI, 1952, 148 f., and the Hermetic literature should be subjected to a more searching disquisition than was done there.

[3] Kore K.42, Nock–Festugière, IV, 13, 16 f. P. Festugière in his introduction, vol. III, cciv f., has not mentioned Empedocles frg. B 107, Diels–Kranz, 5th ed., I, 359. For our reason for doing so cf. K. Kerenyi, *Pythagoras und Orpheus*, 3rd ed., 1950, 22 f.

[4] Kore K.10, Nock–Festugière, IV, 4, 4 f. [5] Cf. e.g. Homer, *Il.* V.426.

[6] A. S. Ferguson in W. Scott, *Hermetica* IV, 1936, 450 n.2, seems to imply this.

[7] This is true of Vergil, *Ecl.* VI.20, 44 f., cf. E. Norden, *Die Geburt des Kindes*, 2nd ed. 1931, 59 f.; Zoroaster, cf. Pliny, *Nat. hist.* VII 72; quoted from Norden, ibid. 65 n.2; and Buddha, cf. W. Weber, *Der Prophet und sein Gott*, Beih. 3 z. Alt., Orient, 1925, 84 n.1, 107.

[8] Kore K.43 f., Nock–Festugière, IV, 14 f. Momus was the personified contumely, and in *PGM* XIII.520, Preisendanz, II, 112, the strong one is Phobos, terror. This is also a related feature in the two myths.

[9] Kore K.62, Nock–Festugière, IV, 20, 14 f.

The two myths, therefore, are built upon the same foundations.[1] Their approach to the religious theme, however, is very different. The magicians regarded their 'chuckling' creator-god with an extreme scepticism.[2] He was not their supreme deity, whoever that may have been. Hermetism, on the other hand, presents us with an essentially optimistic creation myth. It did not altogether pass by the problem of evil; but it limited it to earth and to mankind, and assured its adherents that a divine disciplinary power was sufficient to right the wrongs that were committed on earth. Nowhere did this myth face up to the problem how the infinite could ever touch the finite; and it was therefore easily debased by the wickedness of man on the one hand—for those magicians were truly wicked—and on the other overpowered by human despair, which is the keynote of gnosticism. Further creation myths, particularly the Manichaic myth,[3] might be added; but they might obscure rather than help to clarify the one important point which has to be made as the result of our enquiries so far: it is the utter dissolution of logical argument in all the creation myths which owed their popularity to the recognition of the metaphysical character of the problem of the 'beginning', together with the rejection of Aristotle's metaphysical method in favour of mythology.

Only a short glance has still to be cast upon Hellenistic Judaism. For it is necessary to say a word about Jewish deviations from their firmly established and very holy creation myth. We will not deal with the Jewish influences upon non-Jewish creation myths, but only with Greek, if syncretistic, influences upon Jewish thought. Two convictions have to be considered here: the one is that the world was uncreated, eternal; the other that the world was created, but not by God. The first idea is stated in the pseudo-Philonic treatise 'On the eternity of the world'. It is mainly directed against the Stoic doctrine of conflagration, and there is one argument which is particularly significant here. If, says the author, the cosmos had been created, it would have been infantile not only physically, but also intellectually. This, however, could only be

[1] It may be mentioned that I have found no similarities with the Babylonian creation myths translated by H. Gunkel, *Schöpfung und Chaos*, 1895, 401 f.

[2] I find it hard to believe R. Reitzenstein, *Das iranische Erlösungsmysterium*, 1921, 178, that this god should be Ormuzd.

[3] Cf. W. Bousset, *Hauptprobleme*, 1907, 46 f. There is now a great amount of new material, esp. in Mani, *Kephalaia*, e.g. chap. IV, edd. Polotski–Böhlig, I, 1940, 25 f., with which to check and verify the reports in Hegemonius, *Acta Archelai* VIII, ed. C. H. Beeson, 1906, 11 f., and in the other, inferior sources of Bousset, which would make it impossible for such an attempt to be made in this context.

held by a lunatic.[1] Obviously the matter under discussion here was the Stoic doctrine of micro- and macrocosm. What need was there for such rudeness? Disregarding the heterodoxy of this Jewish author,[2] since even a heretic is not necessarily a boor, his attack seems to have been directed also at groups with whom a logical argument was out of the question. I think they were gnostics, who prepared the way for Manichaism. For in Mani's cosmology the doctrine of micro- and macrocosm held an important place.[3] The second view was introduced by the real Philo. In his cosmology he took a rather ambiguous position. His 'official' attitude regarding the problem of creation is orthodox enough,[4] if fortified with numerous Platonic and Stoic rather than biblical arguments.[5] In his aside remarks, however, the Hellenistic influence showed itself more effective. Not only did he employ astrological arguments to prove that the Sabbath was the 'birthday of the cosmos',[6] he also shared with Hermetic, gnostic, and magic teachers the view that two creations had taken place, the first of an intellectual and the second of an empirical world.[7] It was he also who, by introducing the Platonic doctrine of the demiurge, paved the way for Jewish mystics to maintain that the empirical world had been created by an angel.[8]

The decisive feature in Jewish cosmology, and the force which kept it intact, was its reserve towards the aetiological point of view. God had a purpose with his creation, even if 'the mortal frame' was its last and least honoured part,[9] which the elements would claim after death.[10] The meaning of monotheism, which was emphatically upheld by Philo,[11] was that God was both the 'beginning' and the ending, a doctrine which bore fruits of devotion and purity in human souls:[12] 'So that they who seek

[1] Cf. Ps. Philo, *De aeternitate* 71 f., edd. Cohn–Reiter, VI, 95, 3 f.

[2] Cf. on this matter J. Bernays, *Abh. Akad. Berlin* 1882/3, 16 f.

[3] Cf. Mani, *Kephalaia* I, edd. Polotski–Böhlig, I, 1940, 151.

[4] We mean his *De opificio mundi*, edd. Cohn–Wendland, I, I f.

[5] Cf. E. Bréhier, *Les idées . . . de Philon*, 2nd ed., 1925, 78 f.

[6] I. Heinemann, *Philos griech. und jüd. Bildung*, reprint 1962, 112 f., 121.

[7] Cf. the remark by Wolfson, *Philosophy of the Church Fathers* I, 544, quoted from R. McL. Wilson, *The gnostic problem*, 1958, 243 n.179.

[8] So G. G. Scholem, *Major trends in Jewish mysticism*, 1955, 74, 114. It may be added, however, that intermediaries of the divine creation were not unknown in late Judaism, particularly the word of God, cf. G. Lindeskog, *Studien z. n.t.lichen Schöpfungsgedanken*, 1952, 94, and his hand, Wisdom 11:17.

[9] Philo, *Quis rer. div.* 172, Cohn–Wendland, III. 40, 3 f.

[10] Philo, *De spec. leg.* I.266, Cohn–Wendland, V, 64, 16 f.

[11] Cf. e.g. *De leg. alleg.* II.2/3, Cohn–Wendland I, 90, 7 f., 18 f.

[12] Philo, *De plant.* 77, Cohn–Wendland, II, 149, 2 f.

after the "beginning" of creation should be answered that it is God's goodness and grace which he has bestowed upon his creatures. For God's gift, bounty, and grace are all the things in the cosmos, and the cosmos itself.'[1] This teleological view of creation held Judaism together,[2] and preserved it from that bewilderment and despair which caused the sudden eruption of creation myths among the Hellenized inhabitants of the Empire at the time of Christ.

[1] Philo, *De leg. alleg.* III.78, Cohn–Wendland I, 130, 2 f.

[2] Cf. W. Bousset–Gressmann. *Religion des Judentums*, 3rd ed., 1926, 358 f.

11 The 'beginning' in late Judaism and early Christianity

The previous chapter has led us far into a mythology which during the earliest times of Christianity gave to the people dreams of power as well as the conviction of their spiritual impotence. The temptation which lay in these myths was increased by the pressure of economic distress, and of the yoke of the foreign conqueror on the one hand, and by their apparently great antiquity on the other. The solution of the spiritual problems which faced men of that time could only lie, it was felt, at the 'beginning'. For the conviction attested best by Jamblichus at the end of the third century A.D. that 'the essence of the soul is bodyless, uncreated entirely, and indestructible, having its being and life by itself, self-moving throughout, and the beginning of nature and all motion',[1] however tinged by Neo-Platonism, was held by most thoughtful people already at the time of Christ, and was also preached by the Christian teachers.[2] The first two words of the Fourth Gospel were meant, therefore, to take up such convictions and to align them with the Christian message. This, it seems, was a courageous decision on the part of John.[3] For it is evident that Jewish circles tried to tone down the corresponding first words of Genesis. Already in the pre-Christian period, in the 'Book of Jubilees' 2:2, we find simply 'on the first day',[4] to indicate the commencement of God's creative work. And even the divine command to 'the angel of his countenance', ibid., 1:27, that he should 'write for Moses from the beginning of creation[5] to the time when my sanctuary

[1] Jamblichus in Stob., *Ecl.* II.8.43, ed. Wachsmuth, II, 1884, 173.

[2] Cf. e.g. *Acts of Andrew* i, ed. M. Bonnet, II, 1898, 38, 8 f., 'we are not the offspring of time, afterwards to be dissolved by time; we are not a product of motion made to be again destroyed by itself, nor things of earthly birth, ending therein'. M. R. James, *The apocryphal New Testament*, reprinted 1953, 337, has dated this text, of which he says on page 350, 'There is no doubt that it is a piece of the original Acts', in the midst of the third century. I have, however, shown (*The Apostolic Succession*, 1953, 68) that the *Acts of Andrew* were known already to Origen.

[3] So the author of the Fourth Gospel will be called here for the sake of convenience.

[4] Preserved only in the Ethiopian. I have used E. Littmann's German translation in E. Kautzsch, *Pseudepigraphen*, 1900, 31 f.

[5] The Greek in this case was presumably *ἀπ' ἀρχῆς*, as in Isa. 43:13, and not *ἀπ' ἀρχῆς κτίσεως* which is solely a New Testament expression, cf. Mark 10:6; 2 Pet. 3:4; Mark 13:19; Rev. 3:14, but unknown to the LXX.

will be built among them for all eternity', shows a considerable reserve. Similarly the *Treasure cave*, a Syrian Christian collection of Jewish lore on Genesis, says:[1] 'In the beginning, the first day, holy Sunday the beginning and the first-born of all days', a statement which has not only analogies in the New Testament as well as in early Christian writers,[2] but especially in Philo's description of the Sabbath as 'the birthday of the cosmos'.[3] The interest of these writers was, therefore, concentrated upon the day and the subsequent course of time, rather than upon any concept of 'beginning' as such. From this, and from the scarcity of rabbinic remarks on the 'beginning' in Billerbeck's commentary,[4] the conclusion might be drawn that the Jews did not take a great interest in the Greek idea of ἀρχή. This, however, could well be a mistake of modern exegesis. J. Bernays, one of the great Jewish scholars of the nineteenth century, has suggested another reason, which is that a number of questionable, if not altogether heretical views, in particular that of a creation from nothing, and of the future reduction of the world to nothing, were derived from Gen. 1:1, mainly by Alexandrian Jews and Christians;[5] and it is, therefore, perhaps more likely that we find here the traces of the notorious resistance of orthodox Jewry against Alexandrian doctrines.[6]

A real indifference towards the first two words of the Fourth Gospel, however, has certainly been shown by modern commentators: '*Ἐν ἀρχῇ*', says for instance M. Dods,[7] 'is here relatively to creation used as in Gen. 1:1 and Prov. 8:23', and Sir Edwyn Hoskyns still echoes this somewhat ingenuous remark.[8] This is, of course, a defective rather than an erroneous statement. It becomes a hazardous statement, however, when M. Dods continues: 'The article is absent because ἐν ἀρχῇ is

[1] *Treasure cave* I.3, German translation in E. Riessler, *Altjüdisches Schrifttum*, 1928, 942 f., based upon Bezold.

[2] Cf. Rom. 1:29; Col. 1:15, and Epist. Apost. 17, Hennecke–Schneemelcher, *Neutest. Apokryphen* I, 3rd ed., 1959, 135 and n.3.

[3] Cf. H. Windisch, *Handb. z. N.T.*, Erg. Bd., 1920, 384, commenting on Barn. xv.8, who gives the available material. The recent commentary on Barnabas by P. Prigent, *L'épître de Barnabé*, 1961, does not refer to the Philonic parallels.

[4] Billerbeck, II, 333 f. G. G. Scholem, *Major trends in Jewish mysticism*, 1955, 74, also says of related ideas, esp. the divine qualities for the creation and rulership of the world, that their traces 'in Aggadic literature are few, but they exist'.

[5] J. Bernays, *Abh. Akad. Berlin* 1882/3, 14 f., mentioning in particular Philo, Pantaenus, and Clement of Alexandria.

[6] M. Friedländer, *Zur Entstehungsgesch. d. Christentums*, 1894, 143 f.

[7] *The Expositor's Greek New Testament* I, 1897, 683.

[8] *The Fourth Gospel*, ed. F. Noel Davey, 2nd ed., 1947, 140 f.

virtually an adverbial expression.' There can be no doubt that such a claim may be correctly made in most cases where ἀπ' ἀρχῆς is used in the Septuagint, and in many other Hellenistic writers, and this observation will be of considerable importance here. There is, however, no instance either in the Septuagint or in the New Testament for such a use being made of ἐν ἀρχῇ. And there is also an alternative explanation for the omission of the article in John 1:1. It may mean that the word was so clearly defined in the mind of the writer that it did not require an article. We have seen that this was the case in Greek philosophical tradition. Reference to Col. 1:18 may show that the same could apply to John 1:1, even if in Rev. 3:14, 'the beginning of the creation', ἀρχή has the article. The 'adverbial use' proposed by M. Dods is, therefore, not a very safe basis from which to argue.

In fact, the modern neglect shown to the words 'in the beginning' in John 1:1, seems to be largely due to the lack-lustre present-day conceptions of 'beginning' or 'commencement' when compared with the full-blooded Greek term of ἀρχή. It is true to say that only in modern time the claim made by so many Greek philosophers, that this world has no beginning and no ending, has been generally accepted; and that the so-called translation in the New English Bible, 'when all things began', when there is no Greek evidence at all for the 'all things', is yet another sign for the cavalier treatment given to the first verse of John. A correct translation might have been 'the principle is the word'. This can be shown by the decision of Jerome when he was faced with the apparent Jewish unconcern with the first words of Genesis. The pseudo-Philonic treatise *On the biblical antiquities* simply starts with the words *initium mundi*, in quite the same vein as the New English Bible. This was an established formula in the West, for *Vetus Latina* also begins, *in initio*.[1] Jerome, however, being a theological translator, wrote in both cases, Gen. 1:1 and John 1:1, *in principio*. This change makes it clear that even a western theologian, if he was a theologian, refused to lose sight of the connotation of direction and rulership, which is so prominent in the Greek word ἀρχή.

If then it is admitted that the emphatic position of ἐν ἀρχῇ at the start of the Fourth Gospel is intentional the conclusion is at hand that the author wished to commence his book in the same way as the Torah. This means that he desired to bring to the mind of his readers a creation,

[1] Guide Kisch, *Pseudo-Philo's Liber antiquitatum biblicarum*, 1949, 111, cf. ibid., 15 f. I feel sure, however, that it was *Vetus Latina* which followed the Jewish lead, in view of 4 Ezra 6:1, cf. G. H. Box, *The Ezra Apocalypse*, 1912, 64.

which was presumably meant to be a 'new' creation, but which was nevertheless in accordance with, and not in contrast to, God's first creation.[1] John made indeed strenuous efforts to keep in harmony with the accounts of the first creation, given in Genesis.[2] The question must, however, still be considered open whether John attached the same meaning to the word ἀρχή, which lay behind its use for the translation of Genesis 1:1 by the Septuagint, more than three centuries earlier?[3]

In order to answer this question we may be helped by an examination of 1 John 1:1. For this will make it clear how much the author of this epistle deviated from the non-committal, purely adverbial use of ἀπ' ἀρχῆς which is the rule in the Septuagint. However, this rule was not always kept. There is a hint in one or two Septuagint passages,[4] and particularly in the messianic verses Micah 5:1/2, of what might be implied in such a 'purely adverbial' use. In verse 2, 'whose goings forth are from of old, from everlasting', the Greek translator rendered the Hebrew 'from of old' by ἀπ' ἀρχῆς. It was pre-eminently this text which was in the mind of the writer of 1 John 1:1, when he began his epistle with the words, 'that which was from the beginning'.[5] The reference becomes clear when 1 John 1:1 is compared with the different use in Luke 1:2, 'those who were eye-witnesses from the beginning', where, however, the New English Bible with its 'original eye-witnesses' seems to have been somewhat too original. Luke here meant the course of the earthly ministry of Jesus, so to speak the external history of the Christ,

[1] The rhythmical character of the opening of the Fourth Gospel, John 1:1–18, was stated by K. Meyer, *Der Prolog d. Joh.-Evangeliums*, 1902, 3 f., and before, and R. Bultmann, *Das Joh.-Evangelium*, 10th ed., 1941, 5 f., has rightly adopted this view, claiming, however, that vv. 6–8, which K. Meyer claimed to be rhythmical also, should be bracketed as absolutely prosaic, ibid., 29 n.1. He assumes, cf. p. 3 f., that they belong to a polemic against a sect of disciples of the Baptist. No such sect can be shown to have existed, cf. *Harv. Theol. Rev.*, 1962, 81 f. n.12, and I feel convinced that Bultmann's pre-Johannine Logos-hymn is itself a myth.

[2] Cf. Sir E. Hoskyns, loc. cit., with numerous illustrations.

[3] It is a pity that G. Delling's remarks in *Theol. Wörterb.* I, 1933, 479, on ἀρχή in the LXX should be so short as to become virtually unhelpful. It is unconvincing that the use of the word there should have been altogether, or even mainly, determined by the use of Heb. *rôsh* in the masoretic text. The long article in Schleusner, *Novus Thesaurus* I, 1820, 448 f., gives an idea of the wide range of Hebrew terms which were rendered by the one Greek word.

[4] Cf. esp. Isa. 43:13, and Hab. 1:12.

[5] Similar veiled references in the Fourth Gospel to Mic. 5:2 may be found in John 7:41; 8:44. Cf. also the remark on the birth of Jesus by Sir E. Hoskyns, *The Fourth Gospel*, 2nd ed., 1947, 163 f.

whereas 1 John 1:1 speaks of the spiritual history of him 'whose goings out are from the beginning, from eternity'. Since Isaiah 43:13, and Habakkuk 1:12, clearly use *ἀπ' ἀρχῆς* as a divine predicate, and 1 John 2:13/14 shows the same use, there is a good reason to connect 1 John 1:1 with an Old Testament testimony, which under the circumstances can only be Micah 5:2, even if New Testament scholars seem slow to admit this.[1]

There was, therefore, a special messianic emphasis put even on the adverbial use of *ἀρχή* in the Johannine writings which, however, has its analogy in one contemporary Jewish writing, 1 Enoch.[2] It seems reasonable to apply this result to the use of the word in John 1:1, where the conception of the 'new' creation does not seem to be connected with any ideas of a second coming. The subject there is rather the messianic meaning which has to be attributed to the Logos. This concern is all the more understandable when it is remembered that the term 'logos' played its part in contemporary Judaism, as e.g. in 1 Enoch 6:38, and particularly in connection with the creation of the world.[3] It is perhaps important to utter a warning against the idea of a temporal sequence of old and new creation, in the same sense in which Hellenistic mysticism expected that a new aeon would take over from the old one, or the Stoics that after the conflagration there would follow the rise of a new cosmos. The messianic task of the Logos was rather to reveal the 'new' creation, which indeed is the true creation and exists from the 'beginning', than to start afresh.[4] For the task of the Messiah in the Fourth Gospel is the revelation of the Father to fallen humanity. In this respect a passage from the Greek translation of Ecclesiastes is of great importance. The translation shows signs of a recent age, and may not have been

[1] The words have both a messianic and a Johannine ring, wherever they may appear in 1 John. The closest contact between this epistle and the Fourth Gospel is made in 1 John 3:8 with John 8:34 and 44. H. Windisch, *Handb. z. N.T.* xv, 2nd ed., 1933, 115, has outlined clearly the relation between 1 John 2:14 and John 1:1, and C. H. Dodd, *The Johannine Epistles*, 1946, 50, that between 1 John 2:24 and John 1:1, via John 6:56 f., but neither of them has made a remark about the typically Johannine use of *ἀπ' ἀρχῆς*.

[2] Cf. 1 Enoch 48:3, and E. Sjöberg, *Der Menschensohn im äthiop. Henoch Buch*, Skrifter utg. av Kungl. Vetenskapssamf. i Lund XII, 1946.

[3] Cf. the references in G. Lindeskog, *Studien z. n.t.lichen Schöpfungsgedanken*, 1952, 94.

[4] This, I believe, is the meaning of *Gospel of Thomas* log. 17, cf. W. Till, *Bulletin John Rylands Lib.*, 1959, 454 f. I agree largely with K. Grobel, *New Testament Studies*, 1962, 367 f., in his hesitation to regard every saying in this Gospel as *ipso facto* heretical.

accomplished before the beginning of the Christian era.[1] In Eccles. 3:11 ἀπ' ἀρχῆς is used in a way which is strongly reminiscent of that spurious Orphic fragment from which Goethe seems to have derived Faustus' translation of John 1:1, 'in the beginning was the deed'. The preacher says:

He has made all things[2] beautiful in his time; yet he has given the aeon in their heart, that man may not find the work (ποίημα) which God has made from the beginning even unto the end.

This is clearly a doctrine of God's secret creation, which man, 'having been given the aeon in his heart', cannot perceive. We have come across such a doctrine of a real, eternal world as opposed to the empirical world in the Eleatic school, where Parmenides received a special, objective revelation of such a real world from Dike, the goddess of justice; and we remember the important part which Eleatic doctrines played in the pseudo-Philonic treatise *On the eternity of the world.*[3] However, this was never a mere school doctrine. Already at the end of the sixth century B.C. Theognis, the Greek moralist poet, had expressed similar thoughts;[4] and Philo's description of the first, bodyless creation[5] is a testimony for the close alliance of Greek and Jewish thought in the matter. The most impressive form of this conception is some lines of Aristobulus' re-fashioned Orphic 'Diathekai', the fragment from which Goethe, as we have seen, took his inspiration:[6]

> You cannot see him, before at least here upon earth,
> my son, I have shown you where I find his
> footprints, and the sturdy hand of the strong god.
> 20 Himself, however, I do not see. For a cloud is fixed
> over that which remains, for me and all men in ten tables.[7]

[1] Eccles. 3:11, σὺν τὰ πάντα ἐποίησεν is clearly a forerunner of the systematic translation of the Hebrew *nota accusativi et* by Greek σύν, which is to be found in Aquila, cf. Schleusner, *Novus Thesaurus* V, 1821, 174. This in its turn rested upon the doctrine proposed in Matt. 5:18, which is shown by this example not to have been limited to the Torah, as Billerbeck, I, 246 f., maintains.

[2] The Greek τὰ πάντα is the version of a Hebrew singular, which may show a reserve towards the Eleatic ἓν τὸ πᾶν.

[3] J. Bernays, *Abh. Akad. Berlin*, 1882/3, 12 f.

[4] Cf. H. Ranston, *Ecclesiastes and the early Greek wisdom literature*, 1925, 17 f.

[5] Philo, *De opif. mundi* 29, Cohn–Wendland, I, 9, 4 f.

[6] Orphic. frg. 247, 17 f., ed. O. Kern, *Orphic. fragm.*, 1922, 261.

[7] The ten tablets, originally the ten commandments, are here, I think, already the ten labours of creation, set out in Hagiga 12a, cf. G. H. Box, *The Ezra Apocalypse*, 1912, 83 n.g, and the 'ten qualities with which the world was created', Hagiga 12b, cf. G. G. Scholem, *Major trends in Jewish mysticism*, 1955, 74 n.122.

Orpheus here instructed his son Musaeus in the wisdom of the preacher, 'he has made all things beautiful in his own time', based upon Gen. 1:31; but he also knows that the beginning and the ending are hidden from men's eyes, with the one exception of the Monogenes, the only begotten son.[1] 'No one has ever seen God; but God's only son, he who is nearest to the Father's heart, he has made him known', John 1:18. The Aristobulus fragment is undoubtedly pre-Christian, but we can now establish its 'footprint' in the 'Gospel of truth';[2] we may, therefore, safely state that one of the messianic tasks already in the expectation of Hellenistic Jewry, was the revelation of the real beginning and ending of the world.[3]

All these parallels ought to make it clear how Hellenistic Jewry regarded the first words of Genesis, 'in the beginning', as an urgent challenge to its theology, and particularly urgent because, apart from the last chapters of Job, and of Eccles. 3:11, the Old Testament afforded no clue for its solution. Job 38:1 f. is to be seen in the light of logion 18 of the 'Gospel of Thomas', 'blessed is he who shall stand at the beginning, and he shall know the end, and he shall not taste death'. Such a mystic approach, however, was not open to many, although we find Job's vision compared with 'the end of the Lord' in James 5:11. It will therefore be more instructive here to examine the rationalist approach made by Philo to the problem of the 'beginning'. Philo was without a doubt deeply moved by this problem, as may be seen from the following quotation:[4]

In the beginning God made the heaven and the earth, where the 'beginning' is not, as some believe, to be taken as a beginning in time. For there was no time before the world was made; but it came into being with the cosmos or after

---

[1] Cf. the immediately following verses 22 f. of the 'Diathekai'.

[2] *Gospel of truth*, col. 37, 25, transl. K. Grobel, *The Gospel of truth*, 1960, 176. On p. 177 n.540, Grobel has not noticed the polemics against Orpheus in the words: 'His footprints the will is, and none shall learn of it or cause it to be spied out so that it might be grasped.'

[3] Cf. *Gospel of Thomas*, log. 18, 'for where the beginning is there shall be the end'; *Gospel of truth*, col. 37, 34 f., K. Grobel, 176 f., 'for the father knows the beginning of them all, and their end'; *Kerygma Petri* frg. 2 ed. E. Klostermann, *Apocrypha* 1, 2nd ed., 1908, 19, 'know ye then that there is one god, who has made the beginning of all things, and has the rule over the end etc.' Cf. also how Philo, *De opif. mundi* 54, Cohn–Wendland, 1, 18, 5 f., finds the beginning of all philosophy in the soul's enquiry after the 'beginning'.

[4] Philo, *De opif. mundi* 26 and 28, Cohn–Wendland, 1, 8, 5 f.

it. . . .[1] 28. If, therefore, the 'beginning' is not to be understood as a beginning in time, it seems to signify a beginning according to number, so that 'he made in the beginning' would be equal to 'he made first' the heaven etc.

Philo here eliminated time from his concept of the 'beginning'. Whether or not John did the same depends upon the question whether or not he accepted 'God's own time' in Eccles. 3:11 as an orthodox doctrine. The parallels from Micah 5:2, and 1 John 1:1, make us hesitate to answer the question in the affirmative,[2] because John saw, we believe, the messianic task of revelation and redemption in one. The revelation offered by Christ, consisted for him in the removal of that defilement by sin, which disables men to see God as the Father.[3] Occasional references to 'his time', as in John 7:6 and 13:1, have no connection with Eccles. 3:11. However that may be, it is evident that Philo rejected an earlier, uncritical approach to the problem of the 'beginning' made presumably, but not certainly, by Jews. He characterized the 'beginning' as an ordering principle of natural numbers to which the events of creation were subjected: 'For even if the creator created all things simultaneously', says Philo,[4] 'creation would nevertheless have an order. For nothing is beautiful in disorder. And order is sequence and delimitation of things preceding and following each other, if not physically at least according to the planner's intentions.' It is evident that Philo was here under the influence of Platonic and Pythagorean ideas, although similar conceptions appear in contemporary Jewish apocalyptics.[5] It happened, however, that Philo,

[1] Similar polemics may be heard from Christian quarters, originating from the end of the second century. Representative may be *Acts of Andrew*, quoted *supra*, p. 190, n.2.

[2] It is still *sub judice* whether or not the Johannine Epistles have the same author as the Fourth Gospel. H. Windisch, *Handb. z. N.T.* XV, 2nd ed., 1930, 189 f., has answered the question in a conservative fashion; C. H. Dodd, *The Johannine Epistles*, 1946, XLVII f., has dealt with it in a more critical vein. In recent time no new argument seems to have emerged in the debate. In any case, however, it is certain that 1 John is relevant for the interpretation of John 1:1 as a representative of the 'Johannine school', whether or not it did come from the same author.

[3] There is an important observation by M. de Goedt, *New Testament Studies* VIII, 1962, 142, about 'le Christ comme le révélateur et comme faisant corps, pour ainsi dire, avec l'objet même de sa révélation' in the Fourth Gospel, which serves to illustrate this thesis.

[4] Philo, *De opif. mundi* 28, Cohn–Wendland, I, 8 f.

[5] Cf. *Apocalypse of Abraham* XVII.13, transl. P. Riessler, *Altjüd.Schrifttum*, 1928, 27, 'thou who dissentanglest the confusion in the world, confusion which in this corrupt world proceeds from the wicked as much as from the just. For thou renewest the world of the just.'

adopting these ideas, found himself on one occasion at least in open contrast to established Jewish orthodox convictions; and fell into a bad logical trap.[1]

In order to show the difference between Philonic and Christian, especially Johannine, thought, yet another parallel to John 1:1 f. will have to be brought into play. By this means it is hoped that John's position in the discussion about the meaning of the 'beginning' may be more accurately defined. This parallel is to be found in Paul, Col. 1:16 f.,

> for in him all things were created in the heavens and upon earth, the visible and the invisible, whether they be thrones or dominions or authorities or powers. The universe has been created by him and for him; and he exists before everything, and the universe is held together in him. He is moreover the head of the body, the church. He is the beginning, the first-born from the dead, to be in all things alone supreme.

Admittedly, the closest contact of this passage with the prologue to the Fourth Gospel is in John 1:3; but this fact is significant because it shows how closely this verse, 'the universe was made by him, and not a single thing was created without him', is fitted to John 1.1. For we are thus warned that it would be a mistake to try and understand Col. 1:16 f. without taking into account John 1:1–3.[2] It is true that Paul clearly signifies that Christ is the 'beginning' of the 'new' creation, but does he thereby maintain that he was not also 'in the beginning' of the first creation? And does John 1:1 deny that the Logos is the 'beginning' of the new creation? If we give a negative answer to both these questions, as we must do, our proposition that the 'new' creation is in fact the revelation of God's true creation receives, I think, decisive support.

It is important to realize that the son as the creator, which is also found

[1] Philo, *Leg. alleg.* II.3, Cohn–Wendland, I, 90, 8 f., 'Thus God is ordered according to the One and the monad, or rather the One and the monad according to God. For every number is younger than the cosmos, as time is too, but God is older than the cosmos, and is its maker.' Philo here attempts, rather illogically, to cut loose from Pythagoreanism, thus endangering his elaborate polemics against those opponents who saw the 'beginning' in time.

[2] This mistake was made by E. Lohmeyer, *Die Briefe an die Kolosser und Philemon*, 8th ed., 1930, 63, 'for the word "beginning" transposes into time that which the word "head" means for the organism'. Lohmeyer gives no account of what that may be, and it should not be overlooked that the 'organic theory of the state' has its basis in the doctrine of *corpus Christi*, and not the other way round, cf. *Zeitschr. d. Sav. Stiftung rom.*, 1953, 299 f.; 1954, 25 f. A Jewish parallel is *Test. Zebulun* 9, 'for all that the Lord has made, has one head'.

in Heb. 1:2,[1] is closely related to Hellenistic, Stoic convictions,[2] the ultimate source of which is to be found in Zeno's saying:[3]

God is the maker of the universe and, so to speak, the father of all, collectively as well as that part of him which permeates the universe, which is called by many names according to its different powers.

Josephus quoted this or a similar saying when describing God as 'the Father and genesis of the universe, and creator of things human and divine';[4] and it seems even likely that Eccles. 3:11, which probably dates from the end of the third century B.C.,[5] is dependent upon this Stoic doctrine of creation. In all this it is, of course, understood that the part of God which, as Zeno says, 'permeates the universe', is the Logos. It was at the time of Paul a widely spread Jewish conviction that the Logos was the creator of the cosmos, as we have seen before;[6] and this conviction was common to all the three, John, Paul, and the author of the Epistle to the Hebrews. John 1:1–3; Heb. 1:2, and Col. 1:16 f., come, as a recent author has said,[7] 'from the same complex of ideas regarding their point of departure, structure, and place in the history of religion'.

The parallel from Paul may serve, therefore, to make it clear that ἀρχή, meaning 'principle', could be understood by Jewish thinkers at the beginning of the Christian era in two ways, even if they did not connect it naïvely with the conception of time. It could be understood

[1] If our thesis that the 'new' creation is also the true creation is accepted, the whole discussion as to when Christ received the title of Son, cf. E. Käsemann, *Das wandernde Gottesvolk*, 2nd ed., 1957, 89, falls to the ground, and he appears as the eternal son. Cf. J. G. Davies, *He ascended*, 1958, 65:111.

[2] References to Philonic parallels in H. Windisch, *Handb. z. N.T.*, 3rd ed., 10 f., only illustrate how ready Hellenized Judaism was to accept such doctrines.

[3] Diog. Laert. VII.147, ascribes this statement to Zeno. It is, however, not a literal quotation, and H. v. Arnim, *S.V.F.* II, 305, 17 f., is probably right in ascribing its wording to Chrysippus.

[4] Josephus, *Ant.* VII.380.

[5] This, I believe, has been proved by M. Friedländer, *Griech. Philosophie im A.T.*, 1904, 130 f.

[6] It is curious to observe how soon this doctrine was monopolized by Christianity: Amelius, the Neo-Platonist, would never have written his venomous remark about the Christian Logos-doctrine, Euseb., *Praep. Ev.* XI.19.1, if any trace of it had survived in contemporary Judaism.

[7] S. Schulz in 'Studia Theologica', *T.U.* LXXVIII, 1957, 357. It seems, however, that the author, relying as he does on stylistic and linguistic statistics, has not made a case for his separation of Johannine and non-Johannine elements in John 1:1–18. R. Bultmann, *Das Evangelium des Johannes*, 10th ed., 1941, 1, stating that these verses form a unit, is correct.

first as the ordering principle, as Philo wanted it to be seen; and it could be understood as the creative or, as Greek philosophy would have put it, the causative principle. When dealing with this second approach, we have to enter the still largely uncharted waters of the powers mediating between God and the world according to the Jewish teaching of the time. For Judaism—and magic and gnosticism, both learning from it—attempted at that time with great determination to fill the unbridgeable gulf between God and fallen creation. There were God's glory (*shekhina*, *δόξα*), his spirit (*ruach*, *πνεῦμα*),[1] his wisdom (*achamoth*, *σοφία*), his word (*memrah*, *λόγος*),[2] or the 'angel of God',[3] all cast in the role of such a mediating power. Against this tendency of filling the distance between the Godhead and the cosmos with divine powers or emanations or hypostases of the deity, we see Philo relentlessly waging war, but not even succeeding unequivocally within his own theological system. And it was in the course of this war too that, in maintaining the oneness of God, he took the decision in favour of the numerical, ordering principle of Pythagoreanism as the correct meaning of *ἀρχή*. In doing so, however, he almost came to grief, as we have seen.[4]

Now, whilst the choice which was made by Philo may be justified by the incessant pressure brought to bear upon Judaism by the philosophical polytheism of the Stoics and Platonists,[5] it is all the more remarkable that Paul appears to have had no such misgivings. It is a fact that not only in Paul, but also in all the contemporary Jewish authors whose writings are still known to us, the divine order in God's creation may be highly praised,[6] but the reference to number, which is so prominent in

[1] Cf. the important gnostic texts, Apocr. Joh. 23, 3 f. = Sophia of J. Chr. 84, 6 f., ed. W. Till, *T.U.* IX, 87, 209, and Clem. Alex., *Exc. ex Theod.* VII.2; XXV.2, ed. W. Casey, *Stud. et doc.* I, 1934, 44, 66 f.; 64, 339.

[2] Cf. Justin M., *Dial.* LXI, calling this *δύναμίν τινα ἐξ ἑαυτοῦ λογικήν*, cf. M Friedländer, *Zur Entstehungsgesch. d. Christentums*, 1894, 10 n.4.

[3] J. Daniélou, *La théologie du Judéo-Christianisme*, 1958, 169 f., suggests that the 'most venerable angel' in Hermas, Vis. V etc., was identical with the Logos. This is improbable since the personified Logos is not to be found in Hermas.

[4] Philo, *Leg. alleg.* II.3, cf. *supra*, p. 198, n.1. Cf. also M. Friedländer, op. cit., 11, saying that Alexandrian Judaism insisted upon 'regarding the intermediary power as one with, and inseparable from god'. R. McL. Wilson, *The gnostic problem*, 1958, 37 f., might have emphasized a little more this attitude of Philo and his predecessors.

[5] Cf. Celsus' attack upon the Christian prohibition of eating sacrificial meat, 'Existenz und Ordnung', *Festschr. f. Erik Wolf*, 1962, 161 f.

[6] In addition to *Apoc. of Abraham* 17:13, cf. *supra*, p. 197, n.5, cf. the texts quoted by G. Lindeskog, *Studien z. n.t.lichen Schöpfungs-Gedanken* I, 1952, 94 f., Ps. Solom. XVIII.10 f.; 1 Enoch 2; Test. Naphtali 2.

Philo, is conspicuously absent. All the same, the banishment of books like Ecclesiastes by R. Akiba in the Pharisaic restoration after A.D. 70, makes it clear that the danger was strongly felt in those strictly orthodox circles that suchlike literature might lead to speculations about intermediary powers, and that it was perhaps only the means by which Philo sought to counter this development which met with their disapproval.[1] It is also true to say that Paul's 'Pharisaism' shows itself by the fact that he nowhere has quoted the Preacher; but it is equally true that in Alexandrian as well as in Palestinian Judaism speculations about the intermediary powers, which are alluded to in his Epistles so frequently, continued, and influenced rabbinical Jewish thought right down to medieval times.[2] In all these speculations, however, an understanding of the 'beginning', which was basically different from that of Philo, was presupposed. It was, however, no less Greek than that of Philo, but instead of being Pythagorean, it had its origin in Aristotle, and had received its final form from Stoicism. It was the causative understanding of the 'beginning', for the Stoics maintained that 'the first cause is that which causes motion'.[3] This first cause of the Stoics was evidently identical with their conception of an active ἀρχή, taken over from Aristotle by Zeno, as we have seen before. He had held that—as distinct from the four 'causes' of Aristotle—'there are two principles of the universe: The one is the passive principle, the qualityless substance, matter; and the active is the indwelling Logos, god. He, being eternal, creates everything throughout the whole of matter.'[4] Within the Stoic school this doctrine was handed down unto Seneca in the first century A.D., who changed the reference to matter as a 'cause'. He wrote: 'There are two data in nature out of which everything comes into being, cause and matter. Matter rests inactive, something ready for anything, but going to vanish if it remains unmoved', and continued, 'now we seek for the first, universal cause. This must be simple, for matter too is simple. We ask, what is this cause? It is acting reason—it is God'.[5]

[1] M. Friedländer, *Zur Entstehungsgesch. d. Christentums*, 1894, 13 n.11, with rabbinical evidence. Cf. also 'Studia Theol.', 1955, 86 f.

[2] Dr M. Wallenstein has kindly referred me to Jehuda Halevy (Heb. version by Jehuda-ibn-Tibbon), *Kitab al Khazari*, Engl. version by H. Hirschfeld in 'Semitic Series', 1905, 283, V.20, 'thou mayest discover the causes of their causes, till thou arrivest at the spheres, and finally arrivest at the prime cause (*ha 'illah ha-risho'nah*), that is God'.

[3] Aëtius, *Plac.* XI.7 = H. v. Arnim, *S.V.F.*, 119, 12 f.

[4] Diog. Laert. VII.134 = H. v. Arnim, *S.V.F.* II, 111, 8 f.

[5] Seneca, *Epist.* LXV.2; 11 = H. v. Arnim, *S.V.F.* II, 111, 24 f.; 120, 17 f.

The change from Zeno to Seneca was evidently caused by the intrusion of a matter–spirit dualism; and it is likely that Philo too rejected the Stoic doctrine of the 'beginning' for this very reason,[1] whereas his contemporaries amongst the Jewish theologians accepted it nevertheless without hesitation.[2] So also did John, for the analogy in particular of his 'through him all things came to be; no single thing was created without him', to Seneca's saying is very close. The same can be shown also of Paul's saying in Col. 1:16 f.,[3] as soon as the special character of this passage is clearly recognized. For Paul, who here opposed Colossian mystery practices, employed for this purpose the language of the mysteries.[4] It is also true to say that the mysteries were an almost necessary complement to the pantheistic approach, with which the Stoics attempted to solve the problem of the 'beginning'. Nevertheless, it cannot be denied that Paul's Epistle to the Colossians is firmly founded upon an orthodox Jewish belief. It is only the language of the mysteries of which Paul has made use in a way similar to that of Philo, who 'flirte constamment avec le vocabulaire mystique'.[5] In this way Paul writes, Col. 1:17, 'and he is before the universe, and the universe has its system from him, because the universe is created through him and for him',[6] whilst Seneca simply states, 'everything comes into being from matter and cause'; and when Paul states, 'he is the head of the body . . . who is the beginning', Seneca expresses the same thought as 'the first universal cause is acting reason—that is god'. Paul, like John, has avoided any reference to matter, but apart from that he agrees with Seneca.

We thus find Philo and Paul, the two great leaders of thought among the Hellenized Jews at the beginning of the Christian era, employing Greek philosophy for their interpretation of Gen. 1:1. And this verse, as Philo maintained, was at the root of all philosophy, which in turn was 'the most perfect good to enter human life'.[7] Both of them, Paul as

[1] With regard to Philo's relation to Stoicism, I still prefer the more detailed approach made by H. Leisegang, *Der heilige Geist* I.1, 1919, 58 f., to the somewhat summary statement by H. Wolfson, *Philo* I, 1947, 112, 'in fact, the entire philosophy of Philo may be reconstructed as a criticism of Stoicism'.

[2] The sources for this contention may be found in J. Daniélou, *Théologie du Judéo-Christianisme*, 1958, 216 f., who, however, says nothing about the Stoic roots of the identification of the Christ with λόγος.

[3] It is believed that Rev. 3:14 is an echo of this passage, cf. W. Bousset, *Die Offenbarung*, 6th ed., 1906, 331, and therefore needs no special discussion.

[4] Cf. L. Cerfaux in 'Sacra Pagina', *Bibl. Ephem. Theol. Lovan.* XII–XIII, 1959, 375 f.

[5] L. Cerfaux, op. cit. II, 374.

[6] Here I have transposed the last sentence of verse 16 for clarity's sake.

[7] Philo, *De opif. mundi* 54, ed. Cohn–Wendland, I, 18, 5 f.

well as Philo, seem to have made a clear option as to what should be regarded as the first principle of the world. Philo stated emphatically: 'God is the beginning and beyond of the universe. This doctrine is the basis of piety; this doctrine, implanted in the soul, brings forth the most beautiful fruit of purity'.[1] He could also state in passing: 'The beginning, therefore, of creation is God; and the last, and least honourable, the mortal frame, is the end.'[2] Paul we find agreeing with him in several respects. In spite of the Stoic flavour of the following saying of Paul's, it might have been Philo who wrote, 'for his invisible attributes have been visible ever since the world began to the eye of reason in the works which he has made',[3] but it is of course Paul, Rom. 1:20. He contrasts the 'invisible attributes' of God, which may be seen by the 'eye of reason', with his 'eternal power and deity',[4] so as to stress, just as Philo has done, God's transcendence. In doing so, both Paul and Philo were in the first instance dependent upon the Old Testament, and in particular upon Ps. 101:26 f. (LXX). It is, however, equally true that the Pauline saying—with regard to the Philonic passages this will go without any further comment—can be connected with numerous Hellenistic parallels.[5] All the same, it can be stated with an equal certainty that, so far from separating them from orthodox Judaism, this agreement between Philo and Paul shows them as orthodox Pharisaic Jews.

There was therefore only the one big cleavage between the two in their theology of creation that Philo, concentrating upon the monotheistic principle, found the 'beginning' in number, God being the monad and the One, whilst Paul found it in the supreme cause, and nowhere identified this with God. Philo stated a noetic principle of the world, and Paul an active one. Having established that much, we have finally to decide whether this contrast was a necessary one or only accidental, i.e. whether the two positions of Paul and Philo were

[1] Philo, *De plant.* 77, ed. Cohn–Wendland, II, 149, 2 f.

[2] Philo, *Quis rerum divin.* 172, ed. Cohn–Wendland, III, 40, 3 f.

[3] There should be no doubt about the close connection between Rom. 1:20 and Eccl. 3:11; neither can there be any doubt about their connection with Philo's *κόσμος νοητός* cf. G. Lindeskog, *Studien z. n.t.lichen Schöpfungs-Gedanken* I, 1952, 141, and especially the sources in H. Leisegang, *Der heilige Geist* I.1, 1919, 60 n.1, who has, however, not mentioned any parallel traditions. It may be remarked that *Orph. frg.* 247, 17 f. also belongs to this group.

[4] Paul has carefully kept the two apart, since it is impossible to see God and live; but the translators of the New English Bible, presenting his 'eternal power and deity' as 'attributes', do not share the misgivings of the prophet Isaiah 6:3.

[5] The following three could easily be added to: Ps. Aristotle, *De mundo* VI, 397b, 19 f.; 399b, 10 f.; Cic., *Tusc.* 1.70; Ps. Plutarch, *Plac. philos.* 1.6, 879F.

mutually exclusive from a logical point of view or allowed for a wider range of choices. In putting this question we have to take into account the fact that the Jews from Alexandria, where Platonic and Pythagorean traditions were strong, would be expected to follow such a lead, as he did; whereas the Jews from Tarsus, where Stoic teaching had received a great impulse by the sun of Augustus' favour shining upon the local Stoic celebrity Athenodorus, the son of Sandon,[1] would choose Stoicism.

It is, however, not enough to refer to the accidents of birth when attempting to find the reasons for the choices made respectively by Paul and Philo. It is true that both were fond of their home towns. There is no reason that I can see why Paul should not have said, 'I am a Jew, a Tarsian from Cilicia, a citizen of no mean city' (Acts 21:39) even if we have to assume that the speech in question was invented by Luke.[2] And Philo's outstanding position as the leader of an embassy of the Alexandrian Jews to the Emperor Caligula, a hazardous undertaking in view of the Emperor's well-known brutality and antipathy for the Jews, shows not only the trust which Alexandrian Jewry put in him, but also his devotion for his city. Nevertheless, neither he nor Paul was alone in the decision which they took. In the case of Paul it is quite clear that his decision was that of the great majority of his fellow Jews, and presumably fully established in Jewish theology already when he sat 'at the feet of Gamaliel'. Philo's view was undoubtedly that of a minority, and no sayings of other Jewish leaders can be found to support it. It may be remembered, however, how the Jew Trypho in his dialogue with Justin singled out the Platonic philosophy as the most eligible for a young non-Jew in order to illustrate the fact that Philo was presumably not alone with his choice.[3] For the 'Platonic' philosophy at that time was in the first instance that of the *Timaeus*. It is also hard to visualize Philo as the leader of Alexandrian Jewry whilst he differed profoundly from established orthodoxy in a question of basic importance. Under these circumstances his predilection for the widely spread formula εἷς θεός may give an indication where to look for his allies.[4]

More important than the proof that Philo as well as Paul were not

[1] Cf. H. Böhlig, *Die Geisteskultur von Tarses*, 1913, 115 f.

[2] Cf. M. Dibelius, *Studies in the Acts of the Apostles*, 1956, 7.

[3] Justin M., *Dial.* 8.

[4] It seems a pity that E. Peterson, *Heis Theos*, 276 f., where he discusses the Jewish roots of the formula, has not discussed its Philonic use. Paul uses the formula once, defensively, 1 Cor. 8:6; and Jas. 2:19 may express the anti-Pauline bias of its author. For Philo and his allies the common basis was, of course, the *shma'*.

motivated in their respective theories concerning the meaning of the 'beginning' solely or even primarily by their personal likes or dislikes is the other, to which this book as a whole is dedicated, that the choice before them was an exclusive one. As long as the problem of creation was seen as an ontological and not as a teleological one no other explanation of the meaning of the 'beginning' could be logically proposed than those of Philo and Paul. Yet the consequences of their respective choices are of profound significance. The principle chosen by Philo, the ordering, metaphysical principle, made the empirical world a secondary consideration in the interpretation of the relation between God and man, whereas the active, causative principle stated by Paul made it God's world.[1] Both theories had been developed by the Greek genius; and ever since the days of Paul the Christian theologians have tried in vain to find a way in which they might be reconciled. Logically such attempts are destined to fail; but in the field of religion, as we have seen, the logical argument has not proved the most potent one. Perhaps there is one decisive reason for the Christian why he ought to concentrate upon the Pauline solution. For whilst Philo said, 'the last and least honoured part, the mortal frame, is the end', the Christian seer heard Christ saying, 'I am alpha and omega, the first and the last, the beginning and the end' (Rev. 22:13).

[1] Here, I believe, the root of the deep contrast between Philo and Paul, so convincingly analysed by W. Knox, *St Paul and the Church of Jerusalem*, 1925, 129 f., is to be found.

Index

Index of Biblical References